Electric Telegraph Series

An Illustrated Handbook
to
The Electric Telegraph

I0433618

NUMBER 1 IN THE ELECTRIC TELEGRAPH SERIES

COVER PICTURE
The first message is received by the Submarine Telegraph Company in London, from Paris on the Foy-Breguet instrument, in 1851.

Robert Dodwell

An Illustrated Hand Book
to
The Electric Telegraph

Third Edition

TGR Renascent Books
2017

PUBLICATION HISTORY

AN ILLUSTRATED HANDBOOK
TO THE ELECTRIC TELEGRAPH

ROBERT DODWELL

First Edition
1860
Second Edition
1862

Third edition published by
TGR Renascent Books
27 Springdale Court
Mickleover, Derby DE3 9SW
United Kingdom
2017

www.renascentbooks.co.uk

All rights reserved.
No part of this publication may be reproduced,
stored in a retrieval system, or transmitted,
in any form or by any means, without the
prior permission in writing of the
publisher, or as expressly
permitted by law.

© Publisher's Copyright: TGR Renascent Books 2017

ISBN 978-1-9792525-6-0

CONTENTS

Chapter   *Page*

    List of Figures --- ii
    Publisher's Note -------------------------------------- iv
    Series Editor's Forward ---------------------------- v
    Author's Preface -------------------------------------- 1
1 The Electric Telegraph -------------------------- 3
2 The Galvanic Battery --------------------------- 9
3 Telegraphic Instruments --------------------- 23
4 Bright's Bell Telegraph ---------------------- 35
5 The Line -- 41
6 Submarine Cables and Telegraphic Messages ----- 51
7 House-Top Telegraphs --------------------------- 61
8 Dial Telegraphs ----------------------------------- 71
    Appendix --- 77

LIST OF FIGURES

Figure   Page

1 Facsimile of the second edition 1862 title page xv
2 Static electricity generator 10
3 A wet cell 18
4 A galvanic battery 20
5 Coils with magnetic and indicating needles 24
6 Cooke & Wheatstone needle telegraph instrument .. 27
7 Interior arrangement of the instrument 28
8 Telegraph circuit, London to Manchester 31
9 Telegraph wall clock with time code letters 32
10 Horseshoe magnet 37
11 Electro-magnet 37
12 Bright's Bell Telegraph interior mechanism 38
13 Bright's Bell Telegraph with outer cover removed ... 39
14 Telegraph pole, cross-arms and insulator's 42
15 Plan of a telegraph circuit, London to Liverpool 45
16 Subterranean telegraph wires 49
17 First submarine telegraph cable 52
18 Telegraph message form—sending 54
19 Telegraph message form—receiving 56
20 Henley Magneto Dial Telegraph 72
21 Henley Dial Telegraph interior view 74

Publisher's Note

The international telegram service in Britain, inaugurated by private enterprise in 1845 but soon taken over by the General Post Office (GPO), and latterly by British Telecom (BT), ended in 2003. In the United States the service finished when Western Union sent its last telegram in 2006. As a consequence, most people today, in the age of the Internet, Satellite Communications, Mobile Phones, E-mail, Instant Text Messaging and Fibre-Optic Cable, have no idea that the world was once girdled with thin iron wires strung on poles over thousands of miles of often inhospitable terrain, or that armoured cables lay fathoms deep in abysmal darkness on the bottom of the oceans. It was via these fragile threads that the world once communicated.

The purpose of this Electric Telegraph Series is to publish in new editions some of the many books on telegraphy that first appeared in the Victorian era. Neglected and forgotten, dismissed as no longer relevant, these books are a treasure trove for historians of technology, research students and interested lay persons. The technology and operation of the telegraph very quickly achieved a level of development and sophistication that now seems quite staggering, as a perusal of the books in this series will soon show.

One caveat must be mentioned—the men writing these books were in complete ignorance of the *nature* of electricity, although of course fully conversant with its *effects*. Electricity was often called a "mysterious fluid," by Victorians on the analogy that electricity somehow flowed through a wire like water flows through a pipe. It was not until 1897 that the atom was "split" and J. J. Thomson discovered the electron, the sub-atomic particle that is ultimately responsible for the flow of electricity. It was well into the twentieth century before a coherent theory of electricity was developed and promulgated. Be cautious, therefore, when reading early authors on electricity. Those who have the need should consult professionals or up-to-date text-books on the subject. For the remainder, the books in this series will provide wonderfully readable and easy to understand accounts of electricity, which while not always strictly accurate, nevertheless provide everything needed to understand telegraphy, telegraphic circuits, telegraphic instruments and their ubiquitous power sources—hand-turned generators or wet batteries of exceptional size.

SERIES EDITOR'S FORWARD

The author of this book, Robert Dodwell, was born in 1831 at Vauxhall, London, to Edward and Elizabeth Dodwell. On March 11 of that year, at St Mark's, Kennington, he was christened Robert Valentine. Nothing is known of Robert's early life, but by 1851, aged 20, he was working in Liverpool as a telegraph clerk. He returned to London having accumulated enough savings to marry Christiana Blanche Rosson on 10 October, 1855, at St Mary's Church, Islington.

In 1857 their first child, a son, Henry, was born. The couple would eventually have six children, a daughter, Blanche Marianne, arriving in 1858, a second girl, Ellen Loary, in 1860, and a third, Edith, in 1861. Sadly, it seems that Henry and Edith did not survive long into adulthood, and may even have died as children. Another daughter, Fanny Crichton, was born in 1866, and finally another son, William, in 1890.

Shortly after the birth of their first child, Dodwell moved his family to Manchester, on his appointment as an active and very able District Manager for the Magnetic Telegraph Company (always known to its adherents simply as the Magnetic). The family moved into a semi-rural home on Stamford Road, Bowden, in Cheshire, a small village (it is now part of Greater Manchester), near the river Bollin, just SW of Altrincham. It was an ideal home for the expanding family, and especially so as it was adjacent to the

Manchester, Knutsford & Northwich Railway, which made travelling into Manchester very convenient for Dodwell. He probably welcomed not only the station refreshment rooms, but also the telegraph and post office located on the station.

He had not been long with the Magnetic when the company won the Lancashire & Yorkshire Railway Company's tender to reconstruct the whole of its overhead telegraph system. The Railway had constructed this itself, somewhat piecemeal, in 1852 and it was in a very dilapidated condition. The work was planned by Sir Charles Bright, the distinguished telegraph pioneer and inventor, who was engaged as a consultant engineer. The whole scheme was supervised by Dodwell, involving him in the provision of 8,000 telegraph poles, 20,000 insulators, 200 tons of wire, 185 telegraph instruments and 400 Daniell cells (wet batteries). He not only supervised the workforce, but was also responsible for training 200 new telegraph clerks. The work was completed in September 1859.

Dodwell was certainly prospering at this time and his family was growing. Life at Stamford Road was eased a little for Christiana (affectionately known as Chris) by the engagement of Charlotte Donegan as housekeeper. Also employed as servants were two teenage girls, Rachel Cain and Harriet Wade. Dodwell was meanwhile organising an exhibition of telegraphy in Manchester, in connection with the British Association. He published a circular to advertise it when it opened in 1861.

On 25 January, 1862, Sir Charles Bright retired from his post as engineer-in-chief to the British & Irish Magnetic Telegraph Company (an amalgamation of the Magnetic with the Irish Magnetic Telegraph Company). He was presented with a testimonial in Manchester, organised by Dodwell, comprising a printed address and a candelabrum costing 140 guineas (£147). Dodwell managed to persuade 260

Series Editor's Forward

officers and other employees of the company to contribute this sum. In his after dinner speech, Bright told the assembled guests that when he joined the Magnetic ten years previously, the company had but 40 miles of line in operation and only 20 employees. Now it had 4,000 miles and employed 1,500 people.

Dodwell was always keen to improve the education of working men, and most especially company employees. Accordingly, on 4 October, he met with thirty of the Magnetic's clerks at the George Street Institution in Manchester, to organise a Mutual Improvement Society. The company, Dodwell announced, would rent a meeting room for their use in the offices at Ducie Street goods station. It would be furnished as a reading room, with a library of forty books relating to telegraphy and electricity, as well as copies of all the current technical journals.

Dodwell spent much time during his residence in Manchester lecturing and writing on telegraphy. As he tells us in the preface to this treatise, he journeyed to "...several towns of Lancashire and Yorkshire, to lecture on electricity and the electric telegraph. During these years (he) lectured on this subject nearly two hundred times, and to above twenty thousand people". As an example of these activities, he travelled to Bacup, about 25 miles north of Manchester, to give a lecture at the Bacup Mechanics Institute. He entitled his talk "Two Hours in a Telegraph Office", and illustrated it by showing various items of telegraphic apparatus he had brought with him. The Society of Arts reported in its Journal (3 March 1865), that he had given the same talk to members of the society, who found it most entertaining. Despite his busy schedule during this period, he also found the time to write this book.

In addition to all this, an unlikely activity was revealed in February 1863, when Dodwell took out a patent for "an invention for an improved method of preventing the destruction of plants by insects and other descriptions of animals, and the means for effecting the same". Although details are lacking, it is probable that this involved the application of electricity in one form or another.

This was the year that Dodwell left the employment of the Magnetic to become the engineer to Bonelli's Telegraph Company in Manchester. He constructed a long six-wire circuit between Manchester and Liverpool, which, on such a prime route, seemed to have a clear commercial future. Unfortunately, Dodwell discovered he had made a bad move when Bonelli's ceased trading the following year, probably due to a lack of capital. He next acted as an agent for W.T. Henley's unsuccessful magneto dial telegraph, which regrettably associated him with another failure.

Things began to look up a little when, in July 1864, he was appointed commission agent in Manchester, upon the founding of the Universal Private Telegraph Company, whose headquarters were at 4 Adelaide Street, Charing Cross, London. His task was to acquire new business from customers wanting to rent the company's telegraph lines for their private use. Travelling around Lancashire and Yorkshire, he managed to attract over thirty new rental contracts in the first year of operation. He interested the company in taking over the circuits he had built for Bonelli, and some experimental transmissions were made. However, terms could not be agreed and the scheme was abandoned.

With this experience behind him, Dodwell now set himself up as a telegraph engineer in the private wire business, a potentially lucrative activity in the era of world-wide trading by Manchester's hard-headed cotton magnates and manufacturing entrepreneurs. From premises at 95 Dale Street, Manchester, at the shop of Isaac Wolf, watchmaker, he advertised himself as a Consulting

Series Editor's Forward

Telegraph Engineer. He moved to rented premises of his own in 1865, at 4 Blue Boar Court. Working from these offices he seems to have had a connection with the General Private Telegraph Company, and although not certain, he may have actually founded the firm. Unfortunately for Dodwell, a lot of other people also saw the potential of private circuits, and the numbers of individuals and partnerships, rather than the joint stock concerns mainly centred on London, became a curiosity of Manchester. It was not an environment where success could be guaranteed.

The Dodwell dwelling in Bowden was, at this time, found to be too small to house the growing family, and by 1871 Dodwell had removed them all to 8 Halliwell Terrace in Salford. However, burdened by expenses he could not meet, his fortunes rapidly spiralled downhill and shortly after the move he was declared bankrupt.

The means by which he recovered from this misfortune began in London in 1870, when the Anglo-Continental Telegram Company, 3 Crown Court, Old Broad Street, London, was founded by Richard Wilhelm Otto Rochs and Edward Calley Manico. These individuals had no previous connection with the telegraph industry, but they did have language skills, important because of the nature of their business. They opened an office in Constantinople, in communication with London, and provided daily news telegrams to these centres from Paris, Berlin, Frankfort, Vienna, Amsterdam and Hamburg. Subscribers to the service paid from £10 to £50 per annum, and the company also handled their private messages. However, it was hardly likely that the company could survive for long as a news agency, in direct competition with Paul Julius Reuter, who twenty years before had founded the hugely successful Reuter's News Agency at the London Royal Exchange. In financial trouble after a mere three years of trading, the company failed and both Rochs and Manico were made bankrupt.

Their misfortunes proved to be Dodwell's way back into prosperity. He acquired the failed Anglo-Continental business in October 1873, by means of a private transaction, which probably indicates that his reputation was unsullied by his own bankruptcy, and that he was able to raise the necessary money. He renamed the business the Cable Telegram Company, with an office at 127 Leadenhall Street. He appointed Otto Rochs as manager of a subsidiary, the Oriental Telegram Agency. This new development in Dodwell's life necessitated a move from Manchester back to London, and the family were soon domiciled at 23 Elm Grove, Fulham, Middlesex.

The Oriental successfully expanded their public and mercantile telegram network throughout India, China, Australia, New Zealand, Japan, and Brazil, and eventually covered all of South America, all over a period of four years. In order to correspond by telegraph, on behalf of subscribers, with agents all over Europe, Asia, the Americas and the Antipodes, Dodwell and his associate George Alger devised an abbreviating code. This made possible the shortest possible messages, important since telegrams every-where were charged by the number of words transmitted. This code was eventually published as a book, entitled "The Serial Code", by the Oriental in 1874, as an octavo edition running to some 250 pages.

With his fortunes thus restored, Dodwell was gratified by his election as a member of the Institution of Electrical Engineers, the citation listing him as Dodwell Valentine Dodwell, of Leadenhall Street. However, on the brink of triumph, things were sullied by a court case between Dodwell and the other directors of the Oriental. The cause of the dispute is not clear, but an unlooked for result was the

Series Editor's Forward

failure of the Oriental in May 1876. A new company, the Oriental and American Telegram Company, was formed to take over, managed by Otto Rochs. However, without Dodwell's hand on the helm, that company failed too, in July 1878.

Dodwell spent the next few years working as a freelance electrical engineer and consultant on all aspects of telegraphy. He finally retired with Christiana to a quieter location, 5 Benhill Road, in Sutton, Surry. He died there, aged 73, on 31 March 1904. He was survived by his wife, who died in 1914, aged 79.

A postscript to Dodwell's life and this book may be recorded. A wealthy American, Schuler Skaats Wheeler, was an avid collector of books, pamphlets, catalogues, advertising literature and anything else to do with electrical telegraphy. Among his collection was a copy of this book, "An Illustrated Handbook to the Electric Telegraph" and also Dodwell and Alger's "Serial Code". Remarkably, it also included the circular which Dodwell caused to be printed, to advertise the Manchester exhibition of telegraphy which he organised in 1861. Before Wheeler's death in 1919, he presented his collection to the library of the American Institute of Electrical Engineers, and Dodwell's works reside there still.

Editor's Comments

The editing of this book has been light, confined mainly to breaking up large blocks of text into shorter, more easily readable paragraphs. Numbered chapters have been introduced, absent in the original publication. Footnotes have been retained and a couple more added, where explanation seemed desirable for modern readers, identified by beginning with the words "Editor's Note". Some punctuation has been altered to reflect modern practice.

Otherwise the text is exactly as it came from Dodwell's pen.

A word or two about some points in the following treatise may be in order, to explain things that were commonplace at the time of publication but have now faded from memory. An example is when Dodwell mentions on page 6, "two great Telegraph Companies". As his contemporary readers would have well known, these were the Electric Telegraph Company, the first company in the world formed to provide public access to telegraphy, and invariably known as the "Electric", and the Magnetic Telegraph Company, always known as the "Magnetic". At one time Dodwell's employer, the Magnetic was the Electric's main rival.

On the same page mention is made of Morse's Printing Telegraph, which utilised "long and short" marks to send signals. This, of course, was the famous dot—dash of the Morse Code, still relatively well known today although fast disappearing with the advent of the internet, email and text messaging.

Page 15 mentions that a flash of lightening was estimated to equal the power of the engines of the Great Eastern, and modern readers may not understand the reference. The SS Great Eastern was an iron ship designed by Isambard Kingdom Brunel. She was by far the largest ship ever built at the time of her launch in 1858, and had sail, paddle and screw propulsion. The total power from her steam engines was estimated at 8,000 hp. She was converted from a passenger liner into a cable-laying ship, the only vessel big enough to carry the vast bulk of the transatlantic telegraph cable that was paid out as she steamed across the Atlantic.

A first mention of gutta percha is made on page 48. This substance became of tremendous importance in telegraphy, as an insulator, especially of undersea cables. It is a rubbery latex resin obtain from the Palaquium fruit tree, found in the rain forests of the Malayan Archipelago. The Gutta

Series Editor's Forward

Percha Company was formed early in the nineteenth century in London, at Wharf Road, Islington, to exploit the substance. Soon Victorians were being provided with gutta percha raincoats, boot soles, shoelaces, walking sticks, doorknobs, inkstands, snuff boxes and photograph frames, as well as brooches, earrings, lockets and buttons. Its use in golf balls transformed the sport, but its true potential as a perfect insulator of submarine telegraph cables came when it was found to show no deterioration when submerged in salt water. It is true to say that few other materials have had such a revolutionary impact on the world, and few others have been so quickly forgotten.

The curious title of Chapter 7, "House-Top Telegraphs", on page 61, perhaps deserves some explanation. In 1859 the London District Telegraph Company was formed, with the aim of having a telegraph station no more than five minutes walk away from any household in the metropolis. To save the cost of erecting telegraph poles along city roads, or digging up the pavements to bury underground lines, a scheme was devised whereby the wires would be fixed to chimney pots or poles mounted on roofs, hence the designation House-Top Telegraphs. This necessitated negotiations with every householder, many hundreds of them, to obtain a wayleave for attaching the wires to their premises. The text inserted by Dodwell for Chapter 7 was previously printed in the magazine "All the Year Round", a weekly periodical founded and owned by Charles Dickens. It provides a substantially true but humorous account of the difficulties faced (and overcome) by the London District Telegraph Company in obtaining the requisite permissions. The District was subjected to much public criticism due to its ugly overhead iron wires, suspended from roof tops and leading to a great mass of overhead lines in the City centre.

Many of these roof top circuits were destroyed by the

snow and gales of the so-called Great Storm of 11 January 1866. Posts were torn from their mountings on the tops of the houses in all directions. In Great George Street, Westminster, fallen wires were entwined around lamp posts, and in Regent Street they were hanging from one side of the road to the other, and drivers of (horse-drawn) vehicles had to remove them in order to pass. Similar damage was experienced in the Euston Road, Farringdon Road, and in many of the leading thoroughfares in the city. During the day of the storm, gangs of men were employed in coiling up the fallen wires, while the Company made emergency arrangements to re-establish communications. More than half the overhead lines were brought down and the Company was unable to raise capital for the immense repair bill. In consequence the rate to send a 15 word telegram was doubled, from 6d. to 1s. 0d. (2½d. to 5p.).

Gordon Roberts
Derby, 2017

[ENTERED AT STATIONERS' HALL.]

AN

ILLUSTRATED HAND BOOK

TO

THE ELECTRIC TELEGRAPH,

BY

ROBERT DODWELL,

DISTRICT ENGINEER TO THE MAGNETIC TELEGRAPH COMPANY.

Second Edition.

LONDON:
T. T. LEMARE, PATERNOSTER ROW.
MANCHESTER: JOHN HEYWOOD, DEANSGAT
THE RAILWAY STATIONS.
ROCHDALE: ALDIS AND PEARSON.

Fig. 1
Facsimile of the second edition 1862 title page

AUTHOR'S PREFACE

At different periods, during the last six years, I have accepted the invitations of my friends in several towns of Lancashire and Yorkshire, to lecture on Electricity and the Electric Telegraph. During these years I have lectured on this subject nearly two hundred times, and to above twenty thousand people.

It is at the expressed desire of several of my friends that I present a reflex of my lectures in the present form. This little book contains, after all, but a *sketch* of one of the greatest wonders of the age. But it may have the good fortune to create a desire for further knowledge; and should it succeed in doing so, that desire can easily be gratified by the study of works of standard value, while the ardent Telegraphist may now keep *au courant* with the latest Electrical discoveries by the perusal of the "ELECTRICIAN" — an ably conducted periodical devoted to the progress of, and discoveries in, Telegraphic science.

It is many years since any Hand Book to the Electric Telegraph was published, and the interest of the public, in Telegraphy, once so intense, has, until very lately, cooled down to a state almost bordering upon indifference. But the Telegraph *Soiree*, held in Manchester last September, in connection with the British Association, and the publication of a journal having for its objects the interests and success of Telegraphy, have again roused the public attention to the importance of the subject.

The writer issues this little book in the hope that it may lead some minds to think — some hands to work; and, if that hope is gratified, he will consider himself well rewarded. There is room for many additional labourers in the Electrical field; and this book will have achieved its purpose if it induces new workers to enter it.

R. D.

Bowdon, December 24th, 1861.

1
THE ELECTRIC TELEGRAPH

I'll put a girdle round about the earth in forty minutes.
Midsummer Night's Dream, Act II, Sc. 2.

If it were the custom to select a text for a lecture as well as for a sermon, no words would be more applicable to the subject of this work than those quoted above; and it is scarcely possible to dwell upon them, and not believe that the immortal bard must have seen some experiments in Electricity, for Electricity was known in those days, although its application to the purposes of the Telegraph was not; and with that wonderful foresight, so frequently shown in his writings, placed in the mouth of a little hobgoblin words that had then no meaning, but which would seem to prophesy that, at some future time, a wish or thought should encircle the globe with all the speed of lightning. It may be averred that this is a groundless supposition, but the passage has been used so frequently as an illustration of the working of the Electric Telegraph, and appears so appropriate as an introduction to a book of this description, that no apology is necessary for using it as a starting point in working out our explanation of that wondrous Electricity which, according to the admirable experiments of Professor Wheatstone, travels at a speed of 288,000 miles in a second of time.

It is not intended to treat of the Electric Telegraph in a scientific, but rather in a popular manner, free from all technical terms. We purpose telling a plain story in a plain way, so as to be understood by everyone. There are many whose occupations prevent them from making Electricity a study, and who have not even a knowledge of the *elements* of *Electrical science*. They imagine the Electric Telegraph to be a mystery, so profound that they are content to remain in ignorance of the means by which the motive power is obtained, and the simple method by which it is worked. In consequence, accounts appear from time to time of impressions so absurd, that credulity can scarcely go further. Who has not heard of the old woman hanging her umbrella on the post to go by Telegraph, or of the good old soul at Gateshead, who addressed a new pair of boots to her son in the Crimea, and hung them on the Telegraph wires? The next morning she found an old pair in their place — not an unlikely thing to occur. "God bless the lad", she exclaimed, "that is good of him; I never thought he'd have sent his old ones back to be repaired!" Indeed, it is a common error to suppose that things are sent by the wires; while others are under the impression that they are pulled like bell wires. And again, we read the following:

> "Mother, how do they send messages by those bits of wires without tearing them to pieces?"
> "They send 'em in a fluid state, my dear".

It is to rectify these mistakes that this little work is written, in the hope that it may circulate among those who do not take sufficient interest in the subject to induce them to purchase the expensive but valuable works of more accomplished authors. Those who desire further information will find no difficulty in obtaining books that will enable them to dive as deeply as they wish into the mysteries of Electricity.

The two principal methods of Telegraphic communication, in Great Britain and Ireland, are:

The Electric Telegraph

First, that which was invented by Messrs. Cooke and Wheatstone, (see Figure 6, p. 27) — the Electric Telegraph *par excellence* — the instrument chiefly used — the Electric Telegraph as it was first introduced to the world, and what it is commonly supposed to be; for very few are aware that several systems are in use, — that many more have been tried and thrown aside, and that on the chief lines of Telegraph, erected for the public service, instruments are used quite different in form, and even in the principle upon which the Electric Telegraph was at first designed.

It seems strange to talk of any Electric Telegraph as being old-fashioned, obsolete, a thing of the past; yet it is a fact that many thousands have been spent in the manufacture of Telegraphic apparatus that has had its day, and done good service too, yet is now thrown aside to make room for improvements, and that these improvements have so altered the Telegraphic system as to render it necessary to commence the study of Telegraphy almost *de novo* on the part of anyone who has ceased to be employed in its vast field, for even a few months, no matter how great his previous experience.

But Cooke and Wheatstone's Telegraph still keeps its ground. Superseded on all the principal commercial Telegraph systems, namely, those worked for the public by the Telegraph Companies, it is still in demand for Railway Telegraphs and other lines where simple apparatus is required, as it is generally in the hands of porters, brakesmen, and other inexperienced workmen. It owes this, no doubt, as much to the character it earned in the first stage of Telegraphic operations, as to its simplicity and durability; and it speaks loudly in praise of its inventors, who first brought it out in such a form as to require but very little alteration to adopt it to present wants. It has stood the test of nearly twenty years, and is still regarded by the

Telegraph Clerk as an old friend not to be treated with indignity.

Secondly, it is intended to describe the system being gradually adopted by the Magnetic Telegraph Company, the Bell Telegraph, invented by Sir Charles Bright and his brother, Mr. Edward Bright, an invention that cannot be excelled for economy, simplicity, correctness, or speed. This will introduce the two systems chiefly in vogue, and do justice alike to the two great Telegraph Companies; but it should be mentioned that while the Magnetic Telegraph Company have other patents in use, the Electric Company also employ, to a great extent, Morse's Printing Telegraph. This is an instrument by which the signals are made through the medium of long and short marks, combined so as to form an alphabet, impressed or embossed on a strip of paper by a sharp tooth or indent working on one end of a lever, the opposite end being depressed by the action of Electro-Magnetism.

It would, however, be impossible, in the space allotted to this work, to enumerate all the systems of Telegraph in use; as well might we go back to describe the old semaphore, or the still more ancient methods of signalling by the use of flags, lamps, or bonfires. Our effort is to describe the earliest and latest systems of Magnetic or Electric Telegraphy, that our readers may know what it is and what it does.†

The first question usually put by anyone wanting to know how the Electric Telegraph is worked is: What is Electricity? and it is as often a matter of surprise to be told that no one has yet solved the question. Electricity has been

---

† In the Railway system of the present time, seven different methods of Telegraphing may be seen, viz: By the whistle, pulling of a bell by a cord, flags, coloured lamps, explosive materials, the semaphore, and lastly the Electric Telegraph.

The Electric Telegraph

the careful study of the most eminent men in Europe and America for upwards of a century; to enumerate their names would fill a page; yet the most eminent philosopher of our day (Dr. Faraday) has said:

> "There was a time when I thought I knew something about the matter, but the longer I live, and the more carefully I study the subject, the more convinced I am of my total ignorance of the nature of Electricity".

Such a verdict from such an eminent authority, unsatisfactory as it may be in regard to the settlement of the question, is entitled to our respect; and Electricity must, for the present, remain in the same category with heat, light, and ozone, the value and properties of which we know and admit, the *nature* of which we are ignorant of. We can describe Electricity as a peculiar force, obtained by peculiar means, and producing the most peculiar results; a force different in its effects from any other with which we are acquainted; but beyond this we know but little, and can only say how it is generated, and what results may be obtained by its powerful, silent, and wonderful aid.

Speaking before a Parliamentary committee appointed to enquire into the subject of Submarine Cables, Mr. Varley, the Electrician to the Electric Telegraph Company, gave the following interesting account of a conversation between London and Odessa; he said:

> 'London first called Berlin, inquired about the weather, and Berlin said, — "It is very cold; have you had any frost in London?" 'We said, "No." 'We then asked for St. Petersburg. In about three minutes we were told to call; St. Petersburg answered, and we put the same question to him. He replied that it was very cold, and the snow was so deep that they were using sledges. We then stated that we had messages for Odessa, and asked for direct communication. He put us through to Odessa; we sent

our messages, and he immediately asked us, — "What weather have you in London?" 'We informed him, and put the same question to him. He stated that they had had no frost in Odessa, and that the flowers were in full bloom. Thus we had spoken through Berlin, where it was freezing hard — through St. Petersburg, where it was colder still — back again into a warmer climate, in one continuous uninterrupted chain.'

Surely this excels anything to be found in the Arabian Nights' Entertainments; and yet, those who are occupied daily in working the Electric Telegraph, get by degrees to regard it as a commercial undertaking, and lose sight of its wondrous powers; the excitement indeed being rather when they cannot keep up a communication, that is at times liable to interruption from atmospheric or other causes, than in being able to talk with a person hundreds of miles away. The latter is within the ordinary routine of Telegraphic life, and ceases to be a subject of astonishment. In explaining how this is brought about, we should treat the subject under three heads.

First. — The Galvanic Battery, or means used by which we generate or obtain the Electric fluid.

Secondly. — The instruments used for communicating the Telegraphic signals.

Thirdly. — The line wire or conducting medium between distant stations, whether Overground, Subterranean, or Submarine.

2
THE GALVANIC BATTERY

Electricity derives its name from the Greek word *electron*, signifying Amber, because it was from this substance that its effects were first discovered. Thales, of Miletus, an eminent Greek philosopher, who flourished about six hundred years before the birth of our Saviour, is said to have produced an attracting power in Amber, by rubbing it with a piece of silk or woollen cloth. It is now well known that glass, resin, sealing wax, and many other substances possess a similar power, and that if they are rubbed briskly with an exciting substance, they will attract to them small pieces of paper, the pith of trees, or other equally light objects. This property is named Electricity; but it is not confined either to the amber or to the substances named; it is applied to a power that is known to pervade all things with which we are acquainted, in a degree more or less intense; and researches of comparatively recent date have proved that, though it is unknown and unseen, except by its effects, it is present in every change that takes place, even to the minutest transformation in the animal or vegetable kingdom.

The five principal methods of producing Electricity are by friction, heat, animal organisation, chemical changes, (whether of combination or decomposition) and Magnetism.

Frictional Electricity is, perhaps, best illustrated by the Glass Cylinder Machine, (Figure 2). *A* is a glass cylinder, supported on a framework, and made to revolve by the

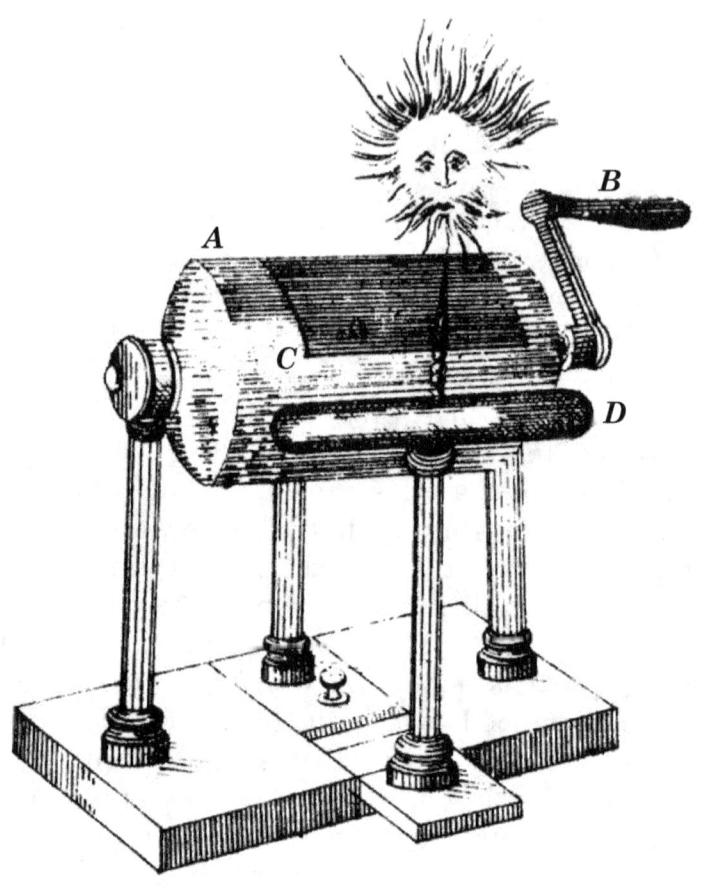

Fig. 2
Static electricity generator, demonstrating how the hair on the head of a wooden doll stands on end when a current is generated.

handle ***B***; ***C*** is a cushion of silk, usually stuffed with horsehair, and having an amalgam or mixture of tin, zinc, and mercury spread upon it to increase the frictional power, when the revolving cylinder is brought in contact with it. ***D*** is a metallic rod, supported on an insulated column. It is called a receiver, because when the cylinder revolves, and the glass is, in consequence, rubbed with some degree of violence against the cushion, the friction causes the Electric fluid to be evolved, that is to say, the following phenomena are manifested. Tiny sparks are seen to fly off from the surface of the glass to the brass points of the receiver; and if then the finger be presented to this receiver ***D***, a zigzag spark of a light bluish colour is seen to pass from it to the finger, while at the same time a slight tingling sensation of pain is felt.

This spark is precisely of the same nature as a flash of lightning, though of course of infinitely less power; and its natural inclination to fly off or discharge itself at the first opportunity, proves the impossibility of making use of Electricity obtained by friction (*static* Electricity, as it is termed) for the purposes of the Electric Telegraph. For if any attempt were made to conduct it along the ordinary Telegraph wires, it would discharge itself at the first bridge, house, or tunnel, if not at the first Telegraph post, into the neighbourhood of which it might come. For this it is not necessary that it should be in positive contact, as it would pass through a considerable distance in order to discharge itself, in preference to passing along the metallic medium.

For Telegraphic purposes, Electricity is required of at least so tenacious a tendency as to prefer a metallic medium to any other, so long as it is not in actual contact with a conductor of Electricity. A familiar instance of frictional Electricity is that of producing a fire by rubbing two dry sticks together. Here we have heat obtained by Electricity.

On the other hand Electricity may be produced by the application of heat; for if two metallic wires, one of silver and the other of platinum, are twisted together, and the ends brought into the neighbourhood of a Magnetic Needle, similar to that which we see suspended over the face of the mariner's compass, the needle will be sensibly disturbed upon the application of the flame of a spirit lamp; thus exhibiting a similar effect to that produced by a current of Electricity obtained from the Galvanic Battery.

The third source of obtaining Electricity, namely, that of animal organisation, will be seen at once to be not available for Telegraphic purposes, and might be dismissed if it did not lead to the contemplation of that wonderful discovery by which is produced Galvanic or Voltaic Electricity. It had long been known that the Gymnotus or Electrical Eel possessed the power of imparting a severe shock to anyone who handled it, or who dipped a finger into the water in which it swam. The Gymnotus is found on the shores of the Mediterranean, and also in the rivers of South America, and possesses a singular formation of the muscles of the body which are perfectly under control; it is in these muscles that the Electricity is concentrated. And Faraday, who conducted a series of experiments upon them, observed, that at the precise moment at which the fish exerted its power, all the water surrounding it became completely charged with Electricity in the same manner as a Leyden jar might be filled from a cylinder machine.

That some animals possess this power to a remarkable extent cannot be doubted, and in none is it so apparent as in the domestic cat; everyone has seen Miss Puss, when attacked by a dog, how she sticks up her back and puffs out her tail; this is, undoubtedly, an effort of Electrical power in the muscles of the animal, and is mechanically produced in the well-known wooden head, the long hair of which stands

The Galvanic Battery

on end when placed on the charged receiver of an Electrical machine, — see Figure 2. Moreover, if a cat be taken into a dark corner, and the fur rubbed briskly with the hand, in a contrary direction to that in which it should lie, brilliant sparks will appear of precisely the same nature and colour as those produced by the machine, proving, beyond a doubt, the intimate connection between Frictional and Animal Electricity. An eminent French writer has recently stated it to be the same power, though in a different degree, that in human beings on every occasion of joy, or grief, or fear, "doth make the hair to stand on end."

Electricity obtained by chemical means, which is called Galvanic or Voltaic, from the two eminent Italian philosophers to whom we are indebted for the discovery, is not altogether to be disconnected from that existing in animals; for it was by the action of metals on the nerves of a frog that this important power was found out. It is said that in or about the year 1789, Galvani was conducting some experiments on some frogs, which, some have thought, were being prepared for soup. It became necessary to hang them on the iron railings in front of his house, and this was done by copper hooks; when he found that the muscles of the frog became violently convulsed, as the wind blew them against the iron railings. On repeating the experiments, he was led to believe that these convulsions were produced by the action of the two metals; and he speedily communicated with his friend Volta, who took up the study only to confirm the discovery that an Electric current could be obtained from the combination of dissimilar metals.

And when the term "current" of Electricity is used, or when it is called the Electric "fluid", it must not be supposed that it runs in a stream large or small; though many analogies have been noticed between wires for Electrical, and pipes for Hydraulic purposes. If it is required to fill a pipe with water, the pipe must be open at one end and closed at the other; but if we want a current or flow of water, — in at one end and out at the other — both ends of the pipe must be open. In the first instance we have the pipe charged with water, in the second a continual flow of water so long as any is poured into the pipe.

So it is with Electricity; a wire or Leyden jar can be filled or, as it is termed, charged, with Electricity; but if it has an opportunity of leaking or escaping, by coming in contact with a conductor, it will continue to flow from the battery in which it is generated through the wire or metallic conductor to the outlet named. Again, no pipe will carry or convey more than a certain quantity of water, according to the size of the bore; if any great pressure is put upon it the pipe will burst, and the water run out or splash back. In the same way, if an extremely powerful current of Electricity is passed through a small wire, the effect will be a recoil or return current, and the wire will first become red hot, and then burst; when, the conducting medium being severed, the battery will assume a quiescent state.

As an instance of this, it may be mentioned that a flash of lightning will, occasionally, strike the Telegraph wires suspended along the Railways, and will traverse them safely till it arrives at some station on the line; here it has to pass through the finer wires, in the coils of the instrument, which are too small to conduct so large a current; and the consequence is that these wires are fused or burnt up, frequently with a loud report, in precisely the same manner that a pipe would burst under a severe pressure of water, or a cannon by a charge of ammunition it is not calculated to bear.

This insufficiency of conducting power, or "resistance," as it is technically termed, it is, that causes the bricks or stones of chimneys to be burst asunder during the passage of atmospheric electricity, termed lightning. Professor Pepper mentions that the mechanical effect of a flash of lightning has been analysed, and it has been found, in one instance, that the power developed in a space of fifty feet was equal to that of a 12,220 horse power engine, or about the power of the engines of the Great Eastern, and that the explosive power was equal to a pressure of three hundred millions of tons.

In order to explain how this Galvanic or Voltaic Electricity is obtained, it is supposed that all are so far acquainted with chemistry as to be aware that all things on the face, and in the interior of the earth, are composed of certain ultimate substances termed Chemical Elements. These were formerly thought to be but four, namely: — earth, fire, air, and water; but these so-called *elements* are not in themselves simple; they are again divisible into upwards of sixty simple and indissoluble bodies. There are many classes, but they are chiefly divided into two groups, "metallic," and "non-metallic." In the first series are found those metals with which we every day have to do, — gold, silver, copper, iron, lead, and a great many others, which are only to be found in the laboratory of the chemist; while in the second series we have oxygen, hydrogen, nitrogen, carbon, sulphur, and many others.†

Again, these bodies or elements are found sometimes in a gaseous, sometimes in a liquid, and sometimes in a solid state; while most of them have a tendency to unite or combine when brought together, and form a distinct compound; or, as it is termed, have an *affinity* for one another. Thus, oxygen when brought into contact with hydrogen, unites with it, and forms water; if brought into

---

† The substance of these remarks on the Chemistry of the Galvanic Battery has been taken by permission of Mr. W. H. Preece, from his contributions to "Our Magazine."

contact with iron, it forms an oxide of iron, or, as it is commonly called "rust", this effect being termed a chemical "combination". Of course, these chemical compounds, such as water or rust, are easily resolved into their constituent elements, and this is called chemical "decomposition".

Again, these elements have greater affinity for one body than another; that is to say, they have a greater tendency to unite with one body than with another; and this affinity is frequently so great that one element, in combination with a second, will release itself in order to unite with a third when introduced to the compound. Thus, if iron be placed in water, the oxygen contained in the water will sever itself from the hydrogen to unite with the iron for which it has a greater affinity, and the oxide of iron or rust is formed while the hydrogen is set free.

Chemical affinity, and Electricity, are so intimately connected, they may be said to be one and the same thing, for the existence of the one is invariably accompanied by the evolution or generation of the other. When iron is dipped into water, the moment that the oxide of iron is formed, Electricity is produced; it, however, requires a peculiar arrangement to detect its presence, and this is effected by the Galvanic Battery, where one metal, having a greater affinity for oxygen, is always brought into connection with a second having a very slight affinity for oxygen, or none at all.

The metals that act upon each other in this manner are placed as follows, the first acting to each succeeding one as copper to zinc: — Platinum, Gold, Silver, Mercury, Copper, Lead, Tin, Iron, Zinc. As the power of a Battery is always in proportion to the difference of the affinity of oxygen for the two metals used, it will be evident that if we had two metals equally oxydizable we should get no current at all, but if on the contrary we select one with a very strong affinity for oxygen, and another with no affinity at all, we obtain the most powerful combination — a current being produced equal in force to the difference of the opposing currents set up by each metal separately. Zinc, therefore, opposed to zinc, or copper opposed to copper, would give no current at all. Zinc opposed to iron would give only a weak current, while zinc opposed to platinum would be the most powerful Battery we could compose. Gold or silver, with zinc, would also be good combinations; but the comparative scarcity and high price of those metals prevent them being used on any scale of a remunerative character; while mercury is unsuitable from its nature; and copper is therefore selected as being cheap, and having but a slight affinity for oxygen, especially when compared with zinc.

These remarks should be carefully re-perused in connection with the following diagram (Figure 3), and its explanation, which we shall render as untechnical as possible. Here is represented a glass vessel in which are two plates, one of copper and one of zinc; the vessel being partially filled with water, and the plates connected together outside by two joined copper wires, from *A* to *B*. A current of Electricity is here evolved; the oxygen of the water, combining with the zinc, forms an oxide of zinc, while the hydrogen settles in bubbles on the copper plate, and is given off freely; but water being a very imperfect conductor of Electricity, it offers a certain resistance to the passage of the Electric current, so much so, that only a slight quantity is excited, while the oxide of zinc formed is insoluble in water, and therefore beyond the first effort no further effect is produced.

To obviate this, namely, in order to render the water a better conductor, and likewise to remove the oxide of zinc, a few drops of sulphuric acid are put in the vessel. This has a great affinity for the oxide of zinc, and, combining with it,

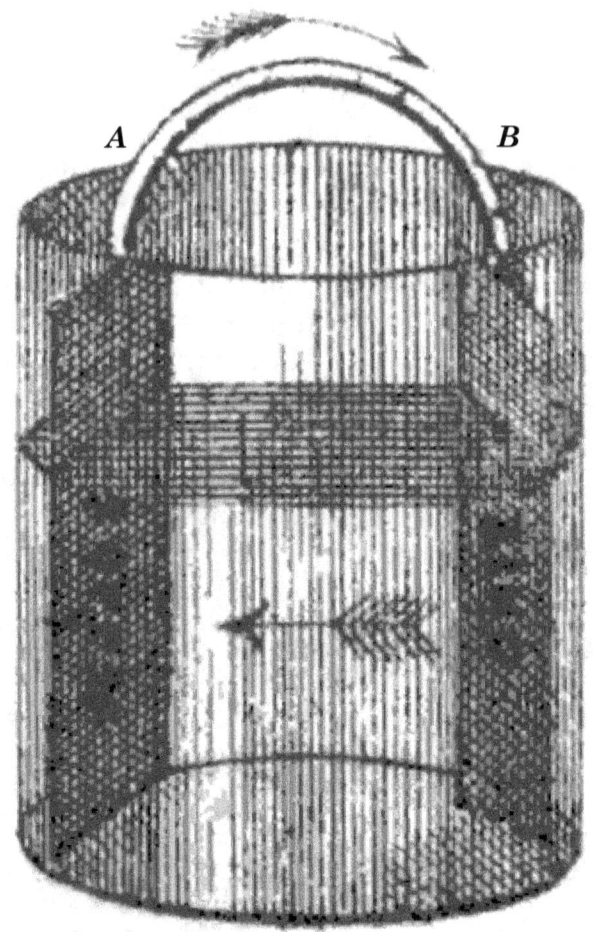

Fig. 3
A wet cell comprising a glass vessel, copper and zinc plates and a solution of sulphuric acid in water

The Galvanic Battery

forms a sulphate of zinc, which *is* soluble in water, and which disappears from the zinc plate, leaving it clean and always exposed to the Galvanic action.

Both Chemical and Electrical action are now being freely evolved in the water, and both may be detected by the senses. On connecting together the two wires, the released oxygen is seen settling in bubbles on the copper plate more rapidly than before; this is the Chemical action, while on holding the two ends of the wires to the tongue, there is felt something of the nature of what is familiarly called a *shock*: this is one of the ways in which the Electrical action manifests itself; and the passage of the invisible agent from the immersed plates, along the wires in the two divergent directions to the point where the wires meet, or into the body which both wires touch, is called an Electric current.

Electricity, like nature, works in circles, and if the conducting wires become broken or detached from either of the plates, the working of the Battery will cease.

As the strength of the Battery is dependent upon the amount of zinc consumed, so also is it necessary to increase its power according to the distance through which the current of Electricity is to pass.

An ordinary whistle can only be heard for a short distance, but a railway whistle can be heard for more than a mile. If one cell or vessel, containing one pair of plates, such as has been described, cannot work an instrument a mile off, six cells can, and if twelve cells will not work an instrument at a distance of 200 miles, 144 will.

In Figure 4, a number of these cells will be seen, the copper plate in the first being connected with the zinc plate in the second cell, and so on to the rest of the series; the power being increased at each successive pair of plates. The first plate, therefore, being a zinc, the last must be a copper termination, and a complete circuit is obtained by uniting

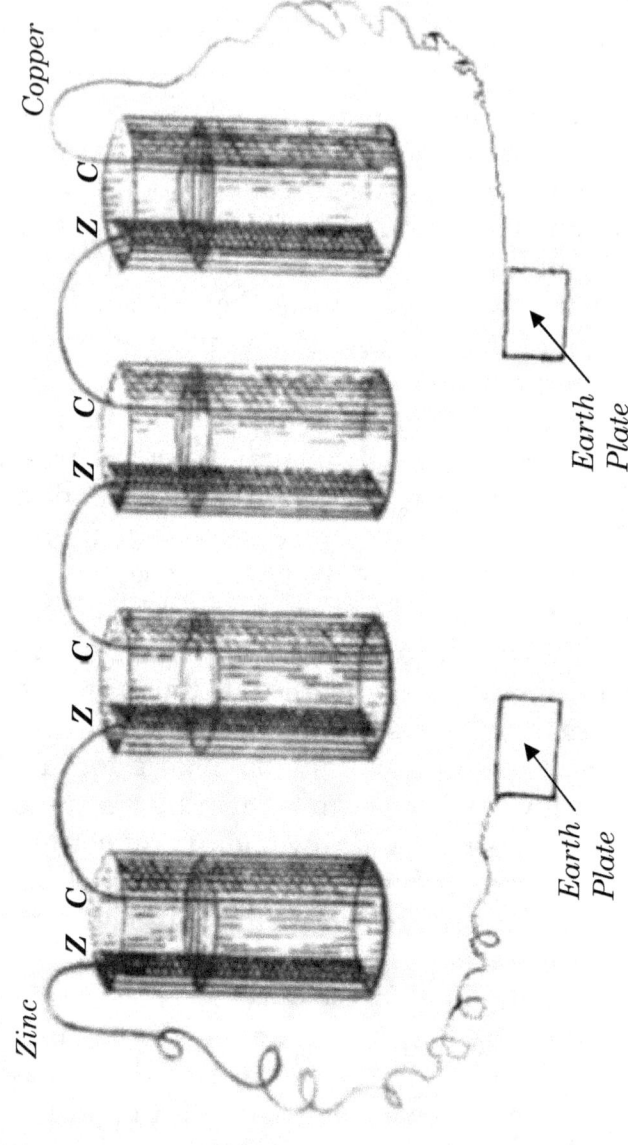

Fig. 4
A Galvanic battery made up by connecting a number of cells with copper wire

The Galvanic Battery

the two terminal plates, zinc and copper, with a piece of copper wire, or some other good metallic conductor. It is on this principle that all Galvanic Batteries are formed; but it is advisable to add that the idea of filling up the cells or vessels with sand which was done at one time in order to avoid the spilling of the acid, has been quite abandoned.

The most favourite form of Battery for Telegraph purposes is that of Professor Daniell, or better still, that designed by Mr. Fuller, the Electrician to Messrs. Silver, of Silvertown. Mr. Varley has very justly remarked that no one, in the present day, would be so absurd as to talk of the "sand" Battery.

It has already been stated that Electricity travels at the speed of 288,000 miles in a second; and if the connecting wires, from each end of the Battery, instead of being carried simply from the zinc to the copper terminals, were to be extended, perfectly insulated, from London to Manchester, the effect would be the same; but it would be necessary either to have both wires so extended in order to complete the circuit, or as will be seen in our illustration (Figure 4), the wires may be carried into the earth. In this case the two opposing currents, from the zinc and copper poles, are disseminated in the earth, and the action of the Battery continues as though the earth itself performed the part of the duplicate wire or return circuit.

(As a familiar instance of Galvanic action, it is well known that porter, if drunk out of a pewter pot, is more welcome than from a stone mug. Here the saliva of the mouth and the porter are two opposite moist conductors, acted upon by the single metallic medium, which is a compound of tin and copper or lead.)

3
TELEGRAPH INSTRUMENTS

Having described the principle upon which the Galvanic Battery is formed, we proceed to the discovery of H. C. Oersted, a Professor of Natural Philosophy at Copenhagen, in or about the year 1819; that if a Magnetic Needle, similar to that used in the Mariner's Compass, were brought into the neighbourhood of a wire, through which a current of Electricity was passing, it would turn from its ordinary course, which is almost due north and south.

This most valuable discovery must be considered as the Keystone to the Electric Telegraph, it being by this method that all the signals have been obtained for Indicating Needle Telegraphs.

It will be better understood upon reference to Figure 5, where *a a* represent two frames of wood, round which is wound from two to four hundred yards of fine copper wire, insulated or coated with silk, or cotton, for its entire length, the ends of which are brought to the two thumbscrews or terminals *c c*; *b* is the indicating needle which works on the same axis as the Magnetic Needle, seen in the interior of the frame; so that any movement of the one necessitates an exactly similar movement of the other.

If wire sufficient to wind round the frames once only was used, and a current of Electricity passed through it, the effect upon the needle would be a minimum; if the wire was wound round twice, double the effect would be obtained; but,

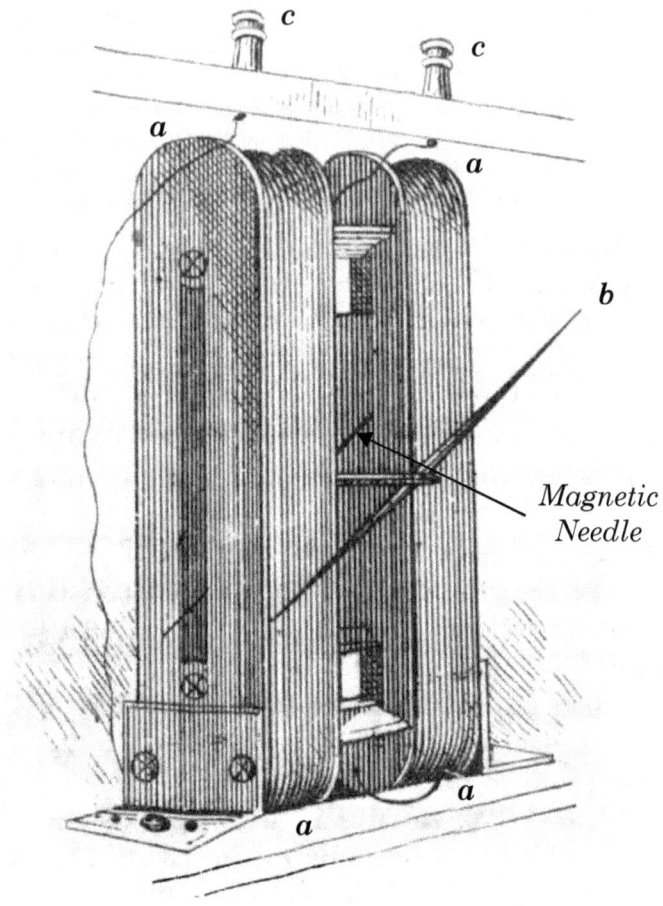

Fig. 5
Coils surrounding a magnetic needle, to which is attached an indicating needle working on the same axis

as it is encircling the frames, within which the Magnetic Needle can move freely, five or six hundred times, the effect upon the needle is increased to a corresponding extent; hence this is termed the *multiplier*, or coil. The necessity for having the wire insulated will be clear, as if it were but simple copper wire it would come in contact when wound round each time, and the current of Electricity, instead of passing completely through the entire length of wire, would take the shortest course it could find.

If the wires from each end of the Battery are brought to the terminals *c c*, the circuit will be completed through the coil, and the needle will be deflected, to the right hand or to the left, according to the direction of the Battery current. This change is effected by placing the wire from the zinc end of the Battery to the right hand terminal of the coils; the wire from the copper termination of the Battery being brought to the remaining coil terminal, or *vice versa*.

The direction in which the needle is deflected is perhaps best remembered by a little device, described by Professor Daniell in his Lectures; supposing that we ourselves are the conducting wires, and the current passes from our head to our heels while we are looking at the Magnetic Needle, the North Pole of the needle will be turned to our right hand. The theory of this disturbance of the Magnetic Needle is, that a current of Electricity proceeds from each end of the Battery; that when the circuit is completed, these currents meeting each other, become opposed, and, being diverted from their direct course, are compelled to move spirally in opposite directions round the wire. In thus moving, or revolving, they act upon the magnetic particles of the needle, producing changes in the positions of those particles, and thus turn the needle from its place; if the poles of the Battery are placed in communication with the earth, an induced current of negative Electricity operates in the same manner with the opposing Battery current.

Following up the simile of the current of water which, passing over a wheel, causes it to revolve, here is seen a current of Electricity passing round a needle which is thus made to move from its position, which position is regained, and the needle becomes stationary, when the current of Electricity is withdrawn, as the water-wheel does when the water ceases to flow.

Who then invented the Electric Telegraph? As this question has frequently given rise to considerable argument, it has been thought advisable to copy the award of Professor Daniell and the late Mr. Brunel, who were appointed arbitrators to decide upon the question; and, it may be well to add, that in this award Mr. Cooke, Professor Wheatstone, and the scientific world generally, fully concurred; it says — "Mr. Cooke is entitled to stand alone as the gentleman to whom this country is indebted for having practically introduced, and carried out, the Electric Telegraph as a useful undertaking, promising to be a work of national importance; and Professor Wheatstone is acknowledged as the scientific man whose profound and successful researches had already prepared the public to receive it as a project capable of practical application."

Mr. William Fothergill Cooke was occupied at Heidelberg, in March, 1836, and saw some experiments with a Galvanic Battery upon the Magnetic Needle. He was so powerfully impressed with the idea, that the influence of the Electric current on the Magnetic Needle might be made use of for Telegraphic signalling, that he at once relinquished his former pursuits, and within a month had contrived an apparatus for the purpose. At this time he was unaware of the exact velocity of the Electric current, and such steps as he had taken were essentially of a very primitive character, but he returned to England, and formed an acquaintance

The letters ETC on the pediment are the initials of the
Electric Telegraph Company

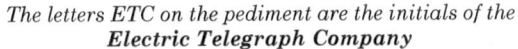

Fig. 6
Front view of a Cooke & Wheatstone needle telegraph instrument

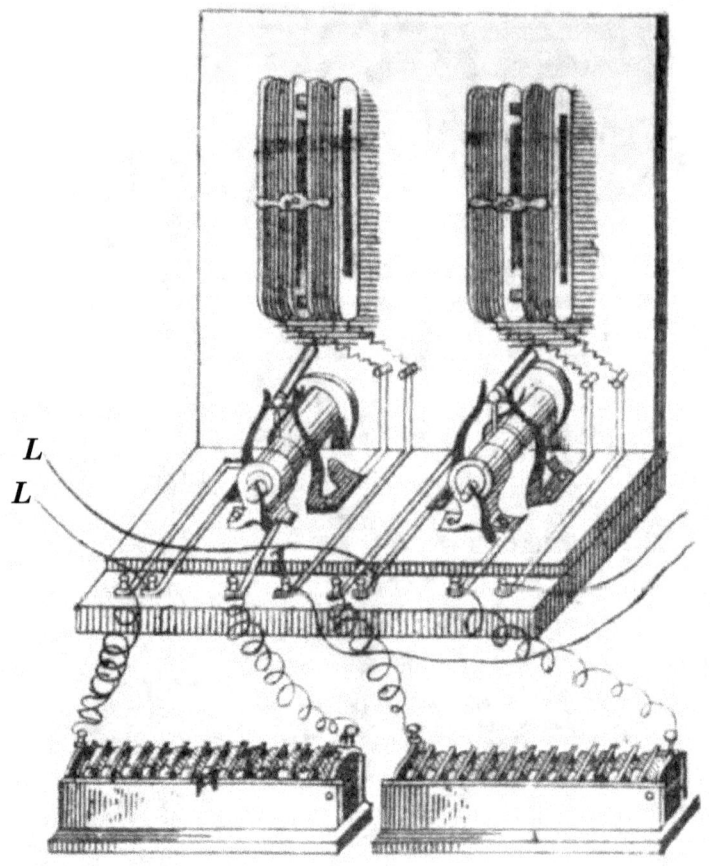

Fig. 7
Interior arrangement of a Cooke & Wheatstone needle telegraph instrument

with Professor Wheatstone, who had already given the subject some attention, especially in arranging a series of keys for opening, closing, and reversing the Electric circuits. The two gentlemen named worked together, and the result was that an instrument was brought out in 1846, so simple and so beautiful in its construction, that for nearly ten years none other was used to so great an extent throughout Great Britain, and it still maintains a very distinguished position notwithstanding the inventions of the last few years.

Figure. 6 gives a front view of this instrument, Figure 7, a sketch of the interior arrangements, and it is most honourably known as COOKE AND WHEATSTONE'S ELECTRIC TELEGRAPH.

On the Dial of Figure 6 will be seen two Indicating Needles, and the alphabet arranged in accordance with the number of signals made use of to denote each letter. There are likewise two handles in front of the instrument, and on reference to Figure 7, the reason for this will be evident, it being throughout a duplicate arrangement. There are two sets of coils; in the centre of each is a Magnetic Needle, corresponding with and working on the same axis with the Indicating Needles in front; these coils being of the same formation as shown in Figure 5. Below the coils are two barrels in connection with the handles on the outside of the instrument, and a peg or tooth on the upper side of these barrels can, by moving the handle to the right hand or to the left, be made to press upon one or the other of the two springs on either side of it. By this means the Galvanic circuit is opened, reversed, or closed, and the Electric current is caused to pass along the wires, through the coils deflecting the magnetised needle, and thence passing out by the wires, marked *L L*, to the line wires, is conducted to the distant station in communication.

Having already remarked that Electricity travels at the

speed of 288,000 miles in a second of time, it is, of course, immaterial to what distance these line wires are carried, or how far away the corresponding instrument may be; the effect is instantaneous, and the current of Electricity passing through the distant coils, at an inappreciable difference of time, causes our own needle, and that of our correspondent, to be moved in the same direction, and at exactly the same moment. The flow of Electricity is stopped, directly the handle is brought to a perpendicular position. With this result the current of Electricity has performed the duty required of it; but, as before stated, it is necessary either to complete the circuit, or find an outlet for it; this may be done either by providing an additional line wire, or better still, and far more economically, by conducting it into the earth as so much waste Electricity.

The alphabet is indicated by the movements of the needles, thus: the left hand needle moved once to the left signifies the "stop," this is given at the end of every word, and likewise means "not understand;" two movements to the left for the letter A, three B; once to the right and then to the left C, the reverse way D, once to the right E, twice F, three times G; the signal E also indicates the signal "understand." The next eight letters are made with the right hand needle in the same way. Once to the left H, twice I, three times K, right and left L, the reverse way M, once to the right N, twice O, thrice P.

The remainder of the letters are made by the combined movements of both needles. Both pointing to the top, or to the letters EH, for Q — both to the bottom and left hand R, twice S, three times T, right and left U — the reverse way V; once to the right W, twice X, three times Y, and the reverse of Q for Z.

Having in this way obtained a signal for every letter of the alphabet, it will be easily understood that words are

Telegraph Instruments

spelt, at full length, letter by letter, the combined movements forming the word complete. All figures are likewise spelt in full, and every word is acknowledged by the signal E (understand), or by "stop," (not understand). In the latter case the word is repeated until the signal E is given, and the clerk who receives it writes it down before another word is proceeded with; this is done by experienced clerks at the rate of from 15 to 20 words a minute.

At one time it was customary to ring a bell, by the agency of the Voltaic current, as a preparatory signal whereby to call the attention of the distant operator. But it was found that the clerks relied too implicitly on the bells, and that serious delays occurred through inattention, whenever they became disordered, which was frequently the case.

Bells are therefore seldom used, at the present time, in connection with Needle Telegraphs; but it is necessary to explain how different stations are informed that a message is to be sent to them. For this purpose we will suppose that Figure 8 represents a line of wire on which instruments are fixed at London, Birmingham, Stafford, and Manchester.

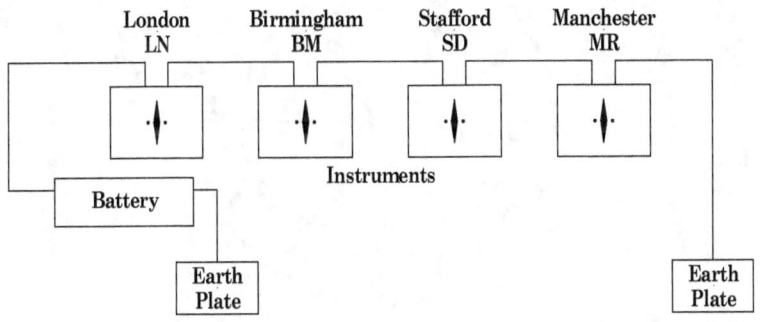

Fig. 8
Telegraph circuit, London to Manchester, with intermediate stations at Birmingham and Stafford

Suppose the Code Signal for London to be **LN**, Birmingham **BM**, Stafford **SD**, and Manchester **MR**, and that London has a message to send to Stafford: the clerk in London will commence pointing to the letters **SD**, and continue to do so till the clerk at Stafford stops him and gives the same signal in reply, London then gives **LN**, and Stafford also repeats that signal; thus he not only knows that it is London who wants his attention, but the same signals are seen at Birmingham and Manchester, though unheeded by the clerks at the latter places, because they are made aware it is **SD** wanted, not **BM** or **MR**. When the message is concluded, if Birmingham or Manchester has a communication for either London or Stafford, he indicates the same by pointing to the letters that stand for the station wanted.

No message is allowed to be sent before another, if handed in after it, and to ensure their transmission in regular order, the fingers of the clock are lettered as shown in Figure 9.

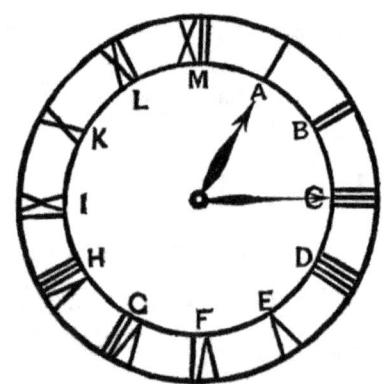

Fig. 9
Telegraph wall clock marked with
time code letters.

Telegraph Instruments

The time is signalled in code, 4-30 being **DF**, 4-35 **DG**, and so on. If Manchester has a message **DF**, and London has one **DG**, the due priority of the message is obtained by these Code Signals being given, so that one message handed in after another cannot be sent before it, but must wait for its turn, according to the time denoted upon it by the Time Code Signal, which is always given at the commencement of the message.

It is impossible to over-rate the value of this beautiful Instrument, and the clerks who first commenced to work the Electric Telegraph will always regard it with a certain degree of affection. It has been the means of forming many friendships; it could scarcely be otherwise, for how can two people converse together year after year, without being interested in one another?

As an instance of this it may be mentioned that in 1851, the year of the Great Exhibition, most of the Provincial clerks went up to London, and almost their first act was to hunt up old chums, whom they had never seen in their lives, but had been talking with, daily, for more than three years. On this occasion they shook hands with men who had flashed, through a distance of more than 200 miles, the cordial invitation, "you'll come and stop with me old chap?" the invite being accepted in a second by the letters **RT**, the telegraphic signal for "all right."

So intimate does this working become, that an experienced clerk will soon detect who is sending to him. The movements of the needles are too slow, too quick, or too slovenly, instead of being regular, and with a dead beat In such a case the clerk may be liable to a fine for any mistake in the message he is receiving, and they will sometimes become so excited as to refuse to take "from such a fellow." Indeed it is not unusual for the Superintendent to have to separate two clerks, because *a few hundred miles is not sufficient to keep them from fighting.*

Mr. Varley states he has noticed that "Telegraph working generally causes great nervous irritation, and the clerks are very prone to quarrel." It should be mentioned that these two statements, which appear so contradictory, are each correct. Quarrelling rarely occurs, however, when the sending and receiving clerks are both experienced in their business; on the contrary, friendship is uniformly maintained; and it may be stated that, since the introduction of young ladies to work the Instruments, Telegraphic courtships have, in more than one instance, led to *hymeneal results.*

The similarity of letters used in signalling by Telegraph, and also the omission of proper punctuation in the messages, hastily written by the public, often cause errors, of the most ludicrous nature, to occur. Thus, a gentleman who had ordered his *gig* to meet him at the station, was understood to require the attendance of his *pig*; while a merchant received an order to ship "forty tons of *fog* to the Mediterranean," the novel cargo having been manufactured from the letters usually accompanying an order to deliver goods *free on board* (fob).

"Your wife has a fine *box* this morning," was intended to intimate the birth of a son; the letters x and y being very similar. "You can have the hundred pieces at sixteen and nine — thousand more at the same rate," was delivered "You can have the hundred pieces at sixteen, and nine thousand more at the same rate," was intended. We don't know if the bargain stood good when the error was discovered.

"Don't come — too late," was a message once sent from Edinburgh to a doctor in London, but being delivered without the pause, it read "Don't come too late," upon which he started off, and only discovered the error after a journey of 400 miles.

4
BRIGHT'S BELL TELEGRAPH

We shall now proceed to explain Bright's Bell Telegraph, which has the advantages of only requiring one line wire, and of communicating by sound instead of by sight. The ear is much more correct than the eye, it is scarcely possible to mistake a sound, but the flickering of a needle before the eye is, at the best of times, indistinct. Then again, as the clerk who receives from a Needle Instrument generally dictates to an amanuensis, sitting by his side, it is found that words are apt to be misunderstood, and are often written down incorrectly.

For instance, a message once sent to Mr. S. R. Graves, was written down for Mr. S. Hargreaves, the error being entirely in the sound, as read off in the dull, monotonous, but generally distinct, tone of the telegraph clerk. The Bell Telegraph avoids this; every letter is sounded out, and written down by the receiving clerk, who never looks at his instrument, so that the eye is not wearied, a great saving is effected in working expenses, and extra security is afforded by the correct transmission of messages.

In order that this instrument should be clearly understood, reference must be made to the discovery of Electro-Magnetism, by which the power of a magnet is given to a piece of soft iron, upon a current of Electricity being passed round it. If an insulated wire is wound round an ordinary kitchen poker, and a current of Electricity, obtained from the Galvanic Battery, is passed through the wire, the poker will become a temporary magnet, possessing all the properties of the Loadstone, even to having a north and south polarity; and this will continue until the current of Electricity is withdrawn, upon which it will cease to have any Magnetic power, and become, what it was before, merely an ordinary kitchen poker.

Figure 10 represents a series of Magnetic Bars screwed together, and forming a Compound Horseshoe Magnet, and the nails attached to it are simply intended to indicate its attractive power.

This same power can be obtained in a piece of soft iron in the manner described above. In Figure 11, *a* represents a bar of soft iron, bent so as to bring the two ends nearer together; round this bar of iron is wound the insulated copper wire, and upon the wires from the Battery being brought to the terminations *b b*, the bar becomes a Magnet, the strength of which depends upon the power of the Battery. If the armature *c* contains a strong hook, it will be found that a considerable weight will be held until the Battery current is withdrawn, when the iron bar will contain no Magnetic power whatever, and the weights will fall away.

On reference to Figure 12, it will be seen how the Messrs. Bright adapted this power to their Bell Telegraph. *A* is the Electro-Magnetic coil: *B* the armature, attracted when a current of Electricity is passed through *A*, and bearing the hammer *C*, which strikes on the alarum *D*; the hammer being released, and held up by a spring, as soon as the Electric current is withdrawn from the coils.

It is impossible to use many words in explaining an instrument so simple, but Figure 13 presents a sketch of the Bell Telegraph complete.

Here are two sets of coils and two Bells, but an

(Continued on page 40)

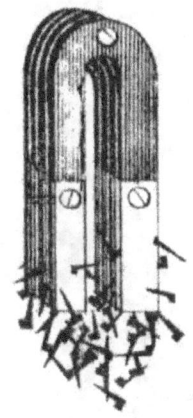

Fig. 10
Horseshoe magnet demonstrating the attractive power of magnetism

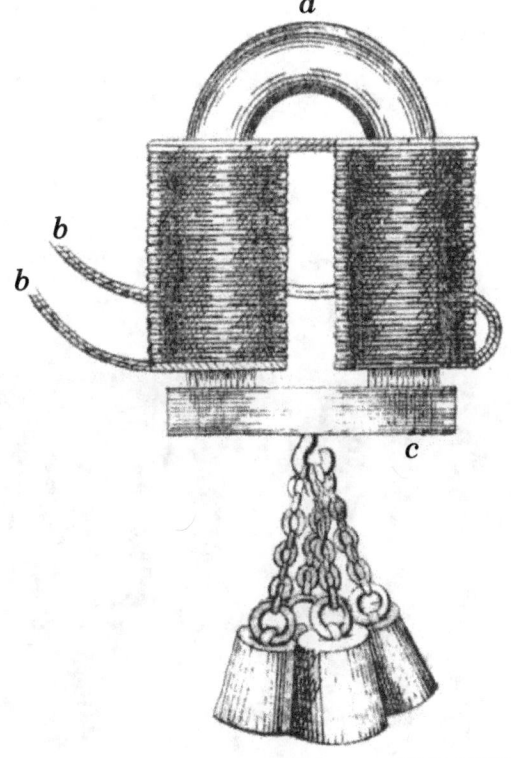

Fig. 11
Unlike a permanent magnet, the electro-magnet *a* only possesses an attractive power if a current is flowing through the coils

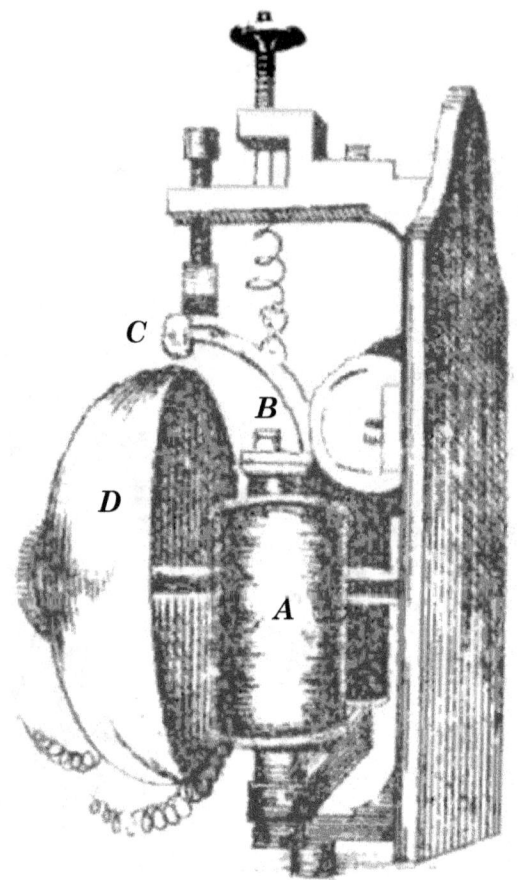

Fig. 12
Bright's Bell Telegraph
Interior mechanism

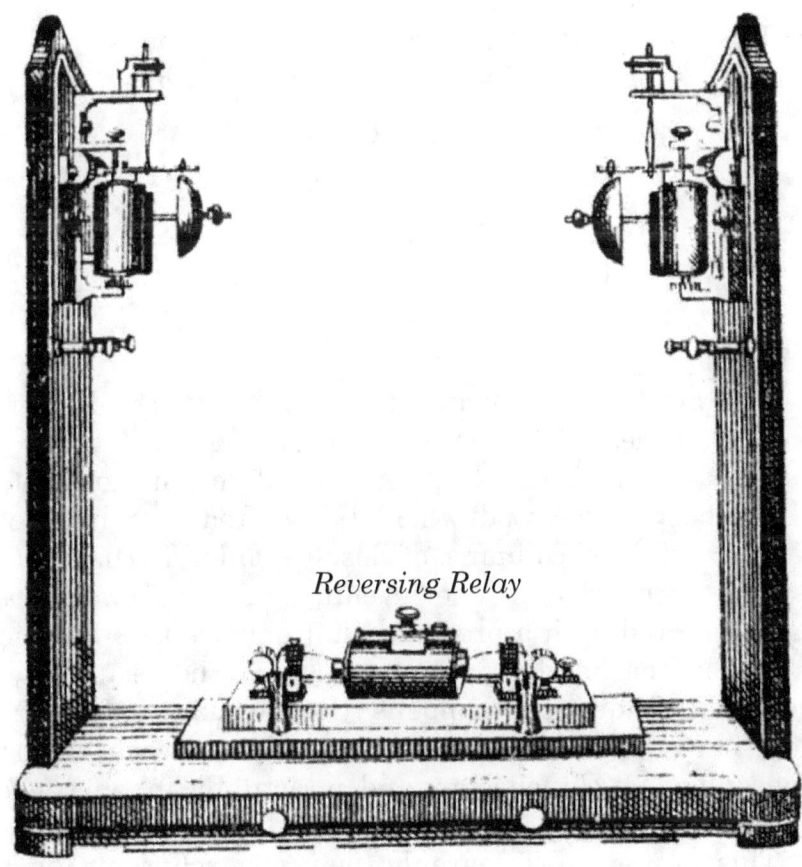

Fig. 13
A complete Bright's Bell Telegraph
with the outer cover removed.

(Continued from page 36)

ingenious piece of mechanism, termed a reversing relay, is brought into action, and instead of using two line wires, the Electro-Magnets are worked, and the Bells rung, by what is termed a Local Battery.

The two small Bells are of different tones, and the alphabet is formed by "ringing the changes," thus:

A	1 right	1 left
B	1 "	2 "
C	1 "	3 "
D	2 "	1 "
E	2 "	
F	2 "	2 "

and so on.

The words are spelt at length, and any signals can be made upon these Bells that may be wanted.

As to the speed and correctness of this instrument, it will be best understood when it is stated that by two of them, six or seven columns, of closely printed Parliamentary News is received in one evening, for the *Manchester Guardian*, and with a precision that enables it to be placed at once in the hands of the compositor. The result is well known; it frequently happens that when a Member is addressing the House, the first part of his speech has been telegraphed to Manchester, and is actually in type before the honourable gentleman has resumed his seat.

This system of Telegraphs is used exclusively by the Magnetic Telegraph Company.

5
THE LINE

This, the third part of our subject, is not the least important part of the Telegraph system, and requires as much care and attention as the power of the Battery, or the manufacture of the Instrument, for without a perfect metallic communication be maintained, free from contact with any other conductor of Electricity, all the efforts of the Signal Clerk, or Electrician, will prove of no avail. This perfect conductor must, therefore, be insulated, so as to afford no escape of the current, especially in wet weather. In different countries dissimilar systems prevail, but, in Great Britain, the Telegraph poles are generally selected from the strongest and straightest larch fir trees; this species of timber being readily obtained of a suitable size and length, and if chosen with care, will be found as lasting as any other sort of wood. After being peeled and planed up, or as it is termed "dressed," they are cut into lengths of from 18 to 30 feet, according to the line for which they are intended, and are then charred or burnt for about six feet at the thickest end, to get rid of any remaining sap which, if left in the wood, would soon cause it to rot after being set in the ground.

In Figure 14, will be seen a cross piece, or "arm" of oak, fixed to the pole by a bolt. This is for the purpose of keeping the wires apart, and to prevent them being blown in contact by the wind, as the poles are from 50 to 60 yards from one another, and if the wires were allowed to hang very closely

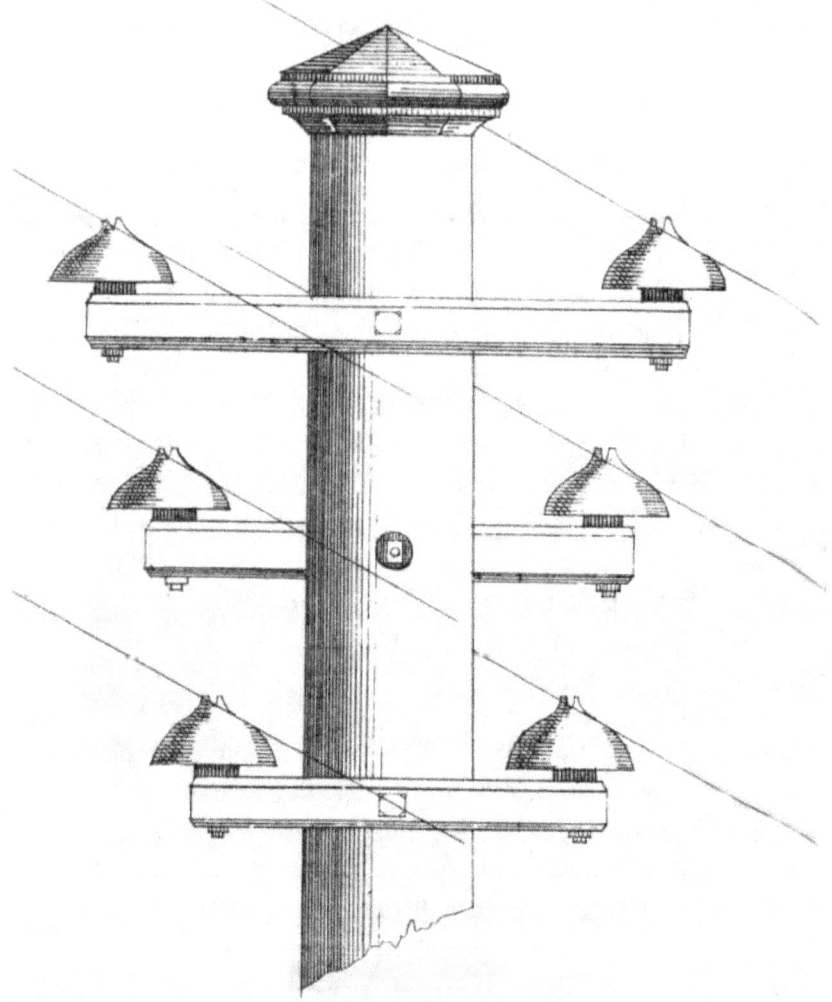

Fig. 14
A telegraph pole with cross-arms and glass insulators,
to the design of Sir Charles Bright

The Line

together, they would, in such a distance, be liable to become twisted, and this would entirely interrupt the due transmission of the Electric Current through each separate wire.

As to the poles and wires, an amusing anecdote is related of two hillbillies who came to a Station, or Depot, in America, where a line of Telegraph had recently been erected.

"Jed," said one, "What are those posts for?"

"To hold up the wires, dude," replied his friend, proud to be able to solve what he considered a scientific problem.

"But what are the wires for?"

This was a temporary downfall for Jed, who, however, soon overcame the difficulty.

"Why, you no-nothing hicky, it's quite clear, the wires are to hold up the posts, darn it."

Damp wood is a conductor of Electricity; if, therefore, the wires rested on the "arms," a certain quantity of Electricity would be lost at each pole whenever a shower of rain came on. The quantity escaping at any one pole would be small, it is true, but when it is considered that the number of poles between London and Glasgow exceeds 10,000, it will be easily understood that the accumulative loss would weaken the power of a current, passing over such a distance, to such an extent as to render it unavailable for the purpose intended on reaching its destination.

To avoid this, a non-conducting substance, termed an Insulator, is fixed on the arm, and it is on this Insulator that the wire is allowed to rest. The Insulators seen in Figure 14 are made of glass, from a design of Sir Charles Bright's; being hollow underneath, namely, in the shape of an umbrella, the rain runs off, leaving the under part dry; and dry glass being a non-conductor of Electricity, so long as this Insulator remains unbroken, the Electrical current will flow safely along the wires in all weathers, and none escapes. This Insulator can be made in Earthenware or Porcelain, but the principle in each case is the same, that of an inverted tube or cup, the underside of which is always dry. Ebonite is now undergoing a trial, and it is anticipated that it will prove one of the best, if not the best, material ever used for insulating overground or aerial lines.

The Line wire is of iron, of No. 8 gauge, Galvanized, that is, coated with zinc, to prevent its being oxidised, for, as in the description of the Battery, it is explained that iron, coming into contact with water, becomes rusted or oxidised, likewise zinc becomes covered with an oxide of zinc. When the wires pass through large towns, the sulphurous gases, given off owing to the large consumption of coal, form a sulphate of zinc; this is quickly washed off, and the iron is soon acted upon in the same way. It, therefore, becomes necessary either to coat the wires with some protecting material, or to put them underground.

A question has often been asked — Why are there so many wires? It has already been explained that each Instrument, or Circuit, requires an independent wire; that is to say, one wire may serve for connecting an Instrument in London with one in Manchester, and this being a direct communication, solely connected with instruments at these two places, no other station can tell what is passing between them. Practically, the work between London and Manchester is found to require several distinct wires, London and Liverpool, and London and Glasgow the same; other wires are placed at the disposal of intermediate places, where the business is not so extensive as that which is carried on between the chief towns. A plan is given in Figure 15, that will show at a glance how an arrangement of eight wires may be made to accommodate a line from London to Birmingham, Manchester and Liverpool; the

The Line

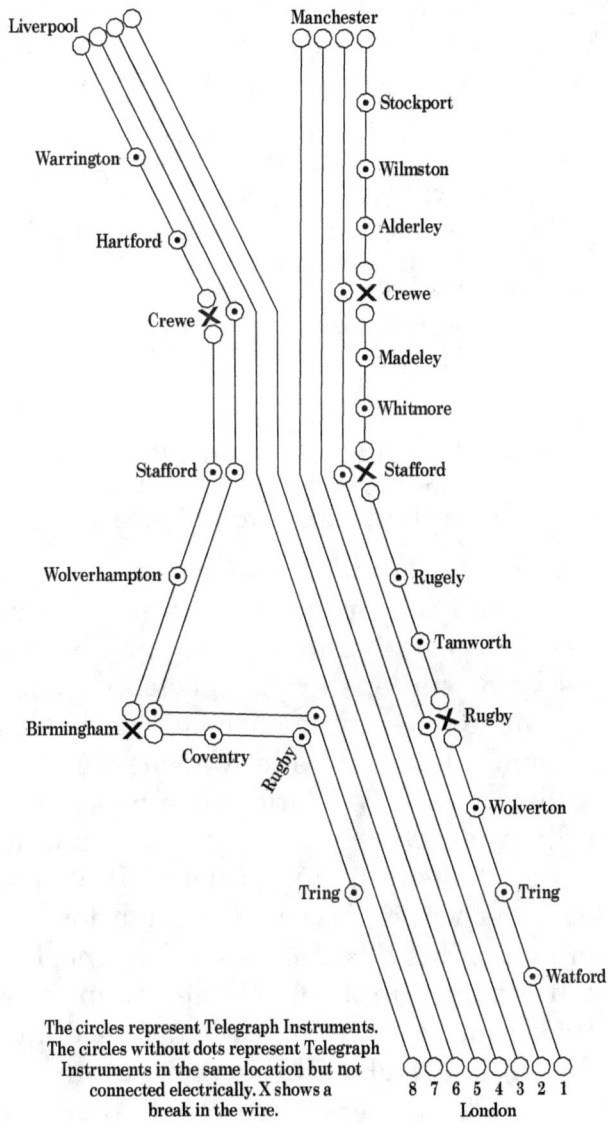

Fig. 15
Plan of a telegraphic circuit comprising eight line wires, linking London to Liverpool and Manchester direct and also via intermediate stations

small circles representing Instruments fixed at the places named. These eight wires are respectively numbered from 1 to 8, and the circuits will be enumerated as follows:

No. 1. — London, Watford, Tring, Wolverton, Rugby.
Rugby, Tamworth, Rugely, Stafford.
Stafford, Whitmore, Madely, Crewe.
Crewe, Alderley, Wilmslow, Stockport, Manchester.
No. 2. — London, Rugby, Stafford, Crewe, Manchester.
No. 3. — London, Manchester.
No. 4. — London, Manchester.
No. 5. — London, Liverpool.
No. 6. — London, Liverpool.
No. 7. — London, Rugby, Birmingham, Stafford, Crewe, Liverpool.
No. 8. — London, Tring, Rugby, Coventry, Birmingham.
Birmingham, Wolverhampton, Stafford, Crewe.
Crewe, Hartford, Warrington, Liverpool.

All stations are brought into communication in this manner, and, united, make a system as perfect as experience enables us to produce it. It must be remembered that when several stations are connected on one wire, the same signals made by any one of them are visible at all the rest; so that while Liverpool and Manchester, for instance, can speak uninterruptedly, Watford can make no movement of the needle that will not be seen simultaneously by London, Tring, Wolverton, and Rugby. If Watford has a message for Liverpool, it is sent from Watford to London, who transmits it on his *direct circuit* to Liverpool.

This will explain, probably, better than in any other way, the reason for so many wires being used; and the odd question also often asked, "Why is there always an uneven number of wires?" is simply answered, there is no such thing. If one wire only is wanted, no more are erected, but as business increases, additional circuits are necessary, and extra wires are added, until, in some places, as many as 22 wires may be counted upon one set of poles.

The Line

It is a common error to suppose that birds are often killed by settling on the wires, and accounted for by the idea that they are struck down by the Electric fluid passing through them. This is not possible, unless they touch two wires, and thus form a connecting medium, through which the Electric current might pass. As they simply settle on the wire, the current continues on its way, preferring the metallic medium to that offered by the body of a bird.

But if a bird had one leg on the wire and the other upon the ground, the Electric fluid might be arrested in its course, and pass, through the body of the bird, into the ground; it is well known, and has been before mentioned, that Electricity invariably takes the shortest route it can find, and it is, therefore, only reasonable to suppose that it would do so in the instance under consideration.

In such a case, if the Electric current was of a sufficiently powerful nature, death would doubtless ensue, and the bird would present the appearance of having been killed by a stroke of lightning. Such an event is, however, not known in Great Britain, owing possibly to the fact that none of our birds possess legs EIGHTEEN FEET LONG!

In India iron rods are used for the conductor instead of wire, not, however, for actual Electrical requirements, but as the Telegraph lines run for miles through the forest and jungle, greater solidity is necessary. The elephants soon tore down the wires, or worse than that, the monkeys turned them into gymnasiums; and it is stated on authority, that on going out one morning, to seek for a breakage, that had interrupted the communication, the Telegraph Superintendent found a mile-and-a-half of ring-tailed monkeys swinging upon his lines, which in consequence had been broken in fifteen different places; so iron rods, supported upon bamboo canes, were substituted for the ordinary Telegraph wire. It is to be hoped that such consideration was duly appreciated by the monkeys.

Circumstances sometimes occur when the iron wires cannot be used as the conductors; this is the case when they are carried through long and damp tunnels, or under the streets: copper wires are therefore used, copper being nine times a better conductor than iron, thus enabling the wires to be reduced to one-ninth the size, a very great consideration when the space is limited, and several wires have to be enclosed in boxing a few inches square.

In order to prevent the escape of the Electric current when passing through the copper wire, the latter is encased in Gutta Percha, an excellent non-conductor of Electricity, if pure, and capable of being drawn solidly on to the wires, so as to avoid any joints, or fissures, which would arise if it were laid on in slips.

Thanks to the elaborate care, and great attention paid to this substance, by the Gutta Percha Company, in sorting it out, and placing it in the hands of their most experienced workmen, for the purpose of being tested, during the process of manufacture, the Electrician has very little now to fear from its use. Subterranean Lines of Telegraph have been condemned owing to the repeated failures of the Insulation, all of which have arisen from the early imperfection of the manufacture of Gutta Percha.

Gutta Percha will remain unchanged for months in air if the light is excluded, and *for years* in water if coated with Stockholm tar.

The best method, therefore, of preparing Gutta Percha Covered Copper Wires, which may be intended for Tunnels, or Subterranean purposes, is to take the number required and wrap them well with tarred yarn. In this form they make up a cable which may be laid in iron, or earthenware, pipes, or as in Figure 16, in wooden troughs. Our illustration presents a section of ten subterranean wires separately

The Line

encased in Gutta Percha wire, to the diameter of 1/4 inch. These are bound together and laid in a wooden box or trough which should then be filled up with sand or tar.

A subterranean line carried out upon such a system would last a considerable time; and has, for its advantages, freedom from atmospheric influences, such as storms of wind and rain; frost, snow, and fog, are equally unavailing in their influence upon it. The cost of maintaining a system where pure Gutta Percha can be relied upon, would, of course, be very slight indeed. No doubt when a fault arises it is not so easily discovered, or repaired, as in aerial lines. There is also a retardation of the current, in wires thus encased, which renders subterranean lines more troublesome to work, in some respects, than those erected overground.

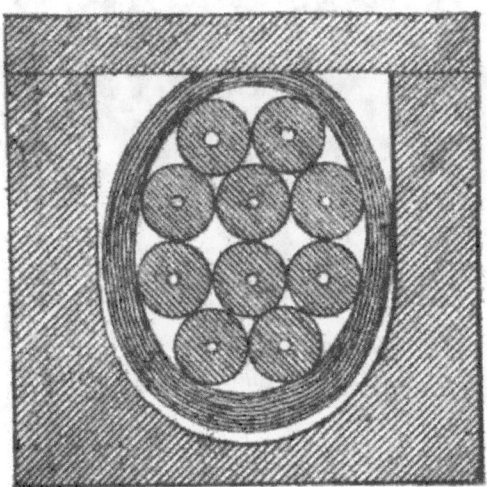

Fig. 16
Ten telegraph wires each encased in gutta percha, bound together and laid in a wooden box filled with sand or tar, preparatory to being buried in the ground.

6
SUBMARINE CABLES AND TELEGRAPHIC MESSAGES

The last part of the subject refers to the Submarine system of Telegraphs, by the aid of which even Europe and America have been united. In Figure 17 will be seen a drawing of the first Submarine Telegraph Cable ever laid. It was on the 17th October, 1851, that this important line was successfully submerged between Dover and Calais, thus connecting the English and French systems of Telegraph. The wire used is similar to that described for Subterranean lines, namely, No. 16 copper wire insulated with a double coating of Gutta Percha, followed by a serving of tarred hemp. Round this tarred rope, ten iron wires, $5/16$ of an inch in diameter, are laid spirally, in order to protect it from abrasion on the undersea rocks. The cost of this Cable was £360 per mile, it was manufactured in three weeks, and it has now been under water for ten years, without any other defect than the occasional damage done by anchors; this is speedily remedied by dragging up the Cable and carefully splicing all the broken parts. When last hauled up it is said to have been as perfect as on the day it was laid. From the success attending this Cable a new Era in Telegraphy commenced, and now sixteen different Cables are carried from the shores of Great Britain, containing an aggregate of forty-four wires, through the medium of which direct communication is maintained daily with Dublin, Paris, Berlin, St. Petersburg, Vienna, &c. &c.

In these pages then, will be found a short description of

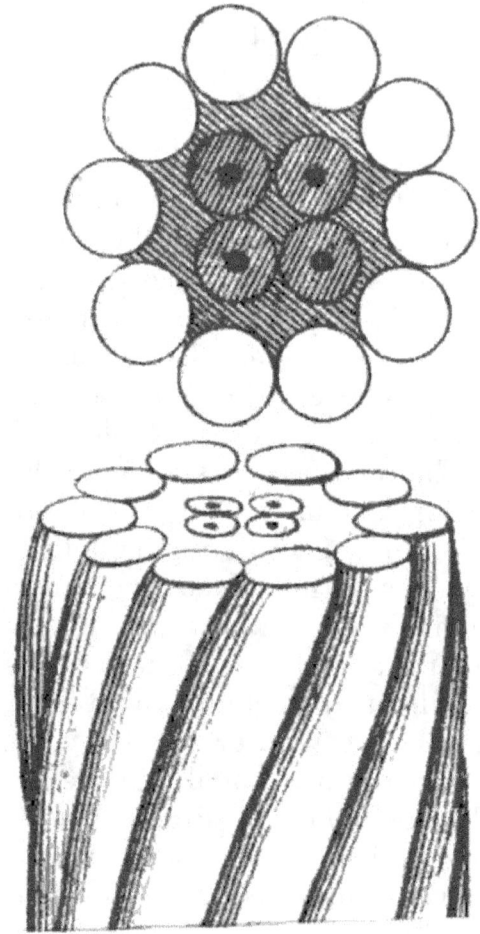

Fig. 17
The first submarine telegraph cable, laid in the
English Channel between Dover and Calais

Submarine Cables and Telegraphic Messages

the working of the Telegraph system in Great Britain and Ireland, — a system that is worked with great success by day and night, in all weathers, be it wet or dry, frost, rain, hail, or snow; it is a servant for the use of the public who may rely upon it, and to enumerate the purposes to which it is applied would be impossible; there is no limit to the subjects contained in the messages that pass through the Telegraph Offices in the course of a few days. Bills, beds, births; murders, marriages; and medicines; news and nurses; rum, races, and robberies; all occupy the time of the Telegraphist, and wonderful as it may appear, it speaks most highly for the honour of the Telegraph service that though these communications pass freely through the hands of the manipulators, the contents have never been known to transpire to anyone save the party to whom the message is addressed. Every message is considered strictly private; for a few minutes it is in the possession of the clerk who sends it, the clerk who receives it, and the superintendent or overlooker appointed to check it, to see that it is correct.

After this, the copy retained is despatched to a Clearing House, to be compared with the original, and to see that the proper charge has been made, and due rapidity observed; and finally, before leaving the Telegraph Company's possession, the paper is carefully destroyed.

That no one may suppose that the transmission of a message is past their comprehension, some explanation of what is required on the part of the public, may be necessary and interesting.

Firstly, we give the form, in Figure 18, in which all messages are required to be written. It is a printed form, supplied gratuitously at the Telegraph Office, to be filled up as denoted in italics, — and it will be well to adopt the following rules in writing all Telegrams, before handing them in at the Telegraph Office:

BRITISH AND IRISH MAGNETIC TELEGRAPH COMPANY,
(LIMITED,)
IN CONNECTION WITH THE SUBMARINE TELEGRAPH COMPANY.

LONDON STATION.

No._____ Date, 18th November, 1861.   No. of Words }_____

Received_____ }
Sent_____ } Code_____

To Station }_____ By me_____ Clerk.

Message   :   :
Porterage :   :
Paid out   :   :

To the Directors,
　　Gentlemen,—I request you to send the following Message according to the conditions printed hereon, and I agree to abide by the same :—

Repetition
Insurance :   :

Total   :   :

From
Johnson,
　　London.

To
Richard Dixon,
　4, *Swan Court,*
　　　Manchester.

Please to write distinctly.

Order a bed for me at the Royal Hotel ;—shall be down at ten forty-five. Meet me.

Signature of } T. JOHNSON.
the Sender.

All Messages taken by this Company are received subject to the following Conditions :

In order to provide against mistakes, and more effectually insure delivery, every Message of consequence ought to be REPEATED, by being sent back from the Station at which it is to be received to the Station from which it is originally sent. Half the usual price for transmission will be charged in addition for repeating the Message. The Company will not be responsible for mistakes or delays in the transmission of, nor the non-delivery of unrepeated Messages, from whatever cause arising. Nor will the Company be responsible for mistakes or delays in the transmission of, nor for the non-delivery of a repeated Message, to any extent above £5, unless it be Insured at the rate of £1 per cent.

NOTICE.—It is expressly provided that the Company cannot Insure or undertake any liability with regard to the transmission of Messages to the Continent of Europe, or elsewhere, beyond the extent of their own lines.

Fig. 18
The form used by the British and Irish Telegraph Company, to be completed preparatory to the sending of a telegraphic message

Submarine Cables and Telegraphic Messages

WRITE PLAINLY — SPELL ALL WORDS IN FULL — USE NO FIGURES IN THE BODY OF THE MESSAGE, THEY COST MORE — PRE-PAY, AND IF YOU WANT AN ANSWER, SAY SO.

This message, when received by the Telegraph Clerk, is counted, the code time attached, the charge calculated, and then, having been prepaid, despatched to the Signals-room, thence to be forwarded to its destination, with all the speed of lightning.

Subjoined are a few instances of singular Telegraph messages that have come under notice.

The following rather obscure message was recently forwarded from one Station master to another:

"A Portmanteau, of the name of H— , was found in a body, with nobody in it. Where shall it be sent to?"

This was intelligible to the party to whom it was directed; but it is surely excelled by Dr. Wilson, who, in his work on Telegraphs, suggests that the insects may have a language of their own, and we are told to imagine the following:

TELEGRAM FROM A SPIDER.
"Fly market light — blue bottles looking up — midges easy — thunder in the air."

In "Prescott's History of the Electric Telegraph," a tale is told of a clerk at Philadelphia, who one day took it into his head to try a joke upon the clerk at New York, who had the reputation of being a steady, matter-of-fact sort of man. Accordingly, he composed and forwarded the following:

Philadelphia, April 1, 1846.
To *Mr. Jones, New York.*
Send me ten dollars at once, so that I can get my clothes.
(Signed) JULIA.
13 words, collect 34 cents.

The clerk at New York, not suspecting any joke, asked

> **BRITISH AND IRISH MAGNETIC TELEGRAPH COMPANY,**
> **(LIMITED,)**
>
> IN EXCLUSIVE CONNECTION WITH THE SUBMARINE TELEGRAPH COMPANY.
>
> MANCHESTER STATION.
>
> Received the following Message the 18th day of November, 1861.
>
> N.B.—You are requested to give no fee or gratuity to the Messenger, and to pay no charges beyond those entered on this sheet.
>
> | From | To | Charges to Pay. |
> |---|---|---|
> | Name.—*Johnson,* | Name.—*Richard Dixon,* | Message ,, ,, |
> | Address.—*London.* | Address.—*4, Swan Court,* | Porterage—*Nil.* |
> | | *Manchester.* | Cab Hire ,, ,, |
> | | | Total ,, ,, |
>
> Order a bed for me at the Royal Hotel;—shall be down at ten forty-five. Meet me.
>
> Signature of the Sender. } T. JOHNSON.
>
> No enquiry about this message can be attended to unless made within ten days of the date of transmission, and upon the production of this paper.
>
> Please sign the Messenger's Ticket, and enter time of delivery.

Fig. 19
The form used by the British and Irish Telegraph Company, on which an incoming message is recorded before it is delivered to the intended recipient

Philadelphia for a better address, to which he replied, "The young lady didn't leave one — look in the Directory for it." The New York operator said he had done so, but as there were over "fifty Jones's in the Directory, he was at a loss to know which one to send it to."

Upon this, the clerk at Philadelphia told him to "send a copy to each of them, and charge 34 cents a-piece." This was done, and several of the messages were paid for; the joke therefore succeeded for the day, but no longer; the Philadelphia gentleman was told he had better not try it again, if he valued his situation.

Secondly, having given the form for a Forwarded Message, we subjoin the form (Figure 19) in which it is received and delivered.

An amusing incident occurred in a Telegraph Office not long ago.

A lady had sent a message to a friend, and had been surprised to see it passed up a shoot, which led to the Signals' Room, at the top of the building. After waiting some time, the answer came down by the same means, and was handed to her in an envelope, duly fastened. She looked wonderingly at it.

"Has this come from London?" she asked.

"Yes, ma'am," replied the clerk.

She paused, and then looking steadily at him, said, "Why, young man, the wafer's wet!" †

He explained that for the sake of secrecy, all messages were enclosed in envelopes before being sent out; but even then she doubted him, and as she left the office he heard her

---

† **Editor's Note**: A "wafer" is a seal (usually red) pasted across the flap of an envelope to show that it has not been opened before reaching the recipient. The young lady in question, noticing that the adhesive was still wet, could not believe the message had just arrived from London, for surely the wafer would have dried during the journey if it had.

say to her companion, "I'm quite sure this is not Louisa's handwriting!"

But the day is going past when such ideas prevail; the iron nerves of the Telegraph are extending most rapidly, and, even in our principal towns, the wires are becoming as familiar as the lamp posts. Fourteen years have elapsed since the Electric Telegraph was first opened formally for the public service; it was to Mr. John Lewis Ricardo, M.P., that the public were mainly indebted for such an advantage. In conjunction with other capitalists, he determined, at a large outlay, to establish a system of Telegraphs throughout the country. The difficulties were immense, not from scientific impediments, but from the distrust and apathy of the public, and especially of the Railway Companies. He, however, persevered, and the names of Cooke, Wheatstone, and Ricardo, will always stand foremost in the History of Telegraphic success. Their claims are undeniable; it is difficult at any time to push a new idea, but this was a system so entirely novel and untried, it required to invent even a new language, and appealed rather to the few, than to the many, for its support; yet, those who were embarked in it laboured to such an extent, that in less than three years over 2000 miles of Telegraph were brought into successful operation.

Mr. Ricardo has stated that such were the difficulties he, at times, experienced, in his attempts to establish the Electric Telegraph Company, that many of his friends left him; and it was only by the persevering encouragement of two or three individuals, and by a large outlay of his own capital, that success was ultimately attained.

But prejudice must cease, the watchword of the present generation is "Onward!" and the latest instance of it is in the adoption of the House Top Telegraph System, an account of which, from "All the Tear Round," follows. In conclusion, we

Submarine Cables and Telegraphic Messages

extract the following from an account of a visit to the printing office of the *Manchester Examiner and Times*, on the 17th August, 1858; it tells of the most wonderful achievement of Telegraphic science: of a success, which, though temporary, is not second in importance as compared with the discoveries of Galvani and Oersted, or the genius of Watt and Stephenson. It was a red-letter day for Telegraphic art, and the Knightly honour bestowed by Her Majesty's command, upon Mr. Charles Bright, was regarded with satisfaction by everyone associated with him in Electrical science. The mind can scarcely realise so great an achievement; for 2500 miles in one continuous, unbroken length, and at a depth of from 50 to 2000 fathoms, this small copper nerve brought Europe and America into instantaneous communication.

Of the risks encountered on shore, the awful storms at sea, the incessant watchful care exhibited, or of the important though temporary results obtained, others have written most ably. The account alluded to runs thus:

> "A breathless boy rushes down and orders the press to be stopped. In an instant he is obeyed, and silence reigns for a moment. What is this piece of news so welcome, that even the sleepy boy's face is lighted up, and the pressmen lean over the galleries, and eagerly listen to hear? And how cheerfully the heavy form is hoisted out from the cylinder, and lowered to the table. The men crowd round as the pressman unlocks the type, and puts in half-a-dozen lines. We could read them by leaning over, but there is no need. They are caught up, and repeated by all around, and in six hours will be Household Words all over Britain.

Here they are:

> Europe and America are united by Telegraph.
> Glory to God in the highest — Peace on Earth,
> and good will toward all men.

> LORDS OF LIGHTNING WE, BY LAND OR WAVE
> THE MYSTIC AGENT SERVES US AS OUR SLAVE.
>
> Henry Schütz-Wilson,
> Assistant Secretary,
> Electric Telegraph Company.

7
HOUSE-TOP TELEGRAPHS

From "All the Year Round."

About twelve years ago, when the tavern fashion of supplying beer and sandwiches at a fixed price became very general, the proprietor of a small suburban pothouse reduced the system to an absurdity by announcing that he sold a glass of ale and an electric shock for four pence. That he really traded in this combination of science and drink is more than doubtful, and his chief object must have been to procure an increase of business by an unusual display of shop-keeping wit. Whatever motive he had to stimulate his humour, the fact should certainly be put upon record that he was a man considerably in advance of his age. He was probably not aware that his philosophy in sport would be made a science in earnest in the space of a few years, any more than many other bold humorists who have been amusing on what they knew nothing about. The period has not yet arrived when the readers of Bishop Wilkins's famous discourse upon aerial navigation will be able to fly to the moon, but the hour is almost at hand when the fanciful announcement of the beer-shop keeper will represent an everyday familiar fact. A glass of ale and an electric shock will shortly be sold for four pence, and the scientific part of the bargain will be something more useful than a mere fillip to the human nerves. It will be an electric shock that sends a message across the house-tops through the web of wires to any one of a hundred and twenty district telegraph stations, that are to be scattered amongst the shopkeepers all over the town.

The industrious spiders have long since formed themselves into a commercial company, called the London District Telegraph Company (Limited), and they have silently, but effectively, spun their trading web. One hundred and sixty miles of wire are now fixed along parapets, through trees, over garrets, round chimney-pots, and across roads on the southern side of the river, and the other one hundred and twenty required miles will soon be fixed in the same manner on the northern side. The difficulty decreases as the work goes on, and the sturdiest Englishman is ready to give up the roof of his castle in the interests of science and the public good, when he finds that many hundreds of his neighbours have already led the way.

The outdoor mechanical exigencies of this London district telegraph require at least six house-top resting-places in the space of a mile. To get these places at the nominal rental of a shilling a year (with three months' notice for removal) has been the object of the company, professedly that a low tariff of charges may be based upon a moderate outlay of capital on the permanent way. The peculiarity of the company's operations, in appealing rather to the public sentiment of the middle and lower classes, than to their sense of business or desire for gain, has prolonged its out-door negotiations; though not to any great extent. The trial may have been severe, but the British householder, with a few exceptions, has nobly stood the test. He has shown that, if properly applied to and properly treated, he may belong to a nation of shopkeepers, and yet be something more than a mere mercenary citizen.

The first time the proposition to electrify all London was brought before the British house holder, it was calculated to inspire considerable alarm. The telegraph, as at present existing, is not a popular institution. Its charges are high;

its working is secret and bewildering to the average mind. Its instrument, as displayed at the railway stations, may look like a mixture of the beer-machine and the eight-day clock; but the curious hieroglyphics and restless arrows on its dial surface are like the differential calculus framed in a gooseberry tart. The unknown may masquerade in the dress of the known; but the railway porter will still shake his head.

When the sole depositary of the telegraph secret has gone to dinner, the whole electric system of that particular railway station must stand absolutely still. A certain amount of familiarity will breed contempt; an equal amount of unfamiliarity will breed awe and dread. The British householder has never seen a voltaic battery kill a cow, but he has heard that it is quite capable of such a feat. The telegraph is worked, in most cases, by a powerful voltaic battery, and therefore the British householder, having a general dread of lightning, logically keeps clear of all such machines.

The British householder (number one) took time to consider. The pole that the company wished to raise upon his roof might not be ornamental; might not suit the taste of his wife, who, at that moment, was unwell; might not meet with the approbation of his landlord, who was very fastidious, and very old. If the company would like to communicate with his landlord, that gentleman was to be found in Berkshire, if he had not gone to Switzerland, if he was not up the Rhine.

The British householder (number sixty) was only one of a firm, and he could give no definite answer without his partners' consent.

The British householder (number sixty-eight) was of a vacillating disposition, and after he had said yes, he took the trouble to run up the street, because he had suddenly decided to say no.

The British householder (number seventy) was the second-mate of a trading vessel, at that time supposed to be running along the South American coast. His wife was not prepared to say whether he had any objection to a flagstaff (although she thought he had not), and she could give no permission to the company until his return.

The British householder (number seventy-four) very politely allowed the survey of his roof; and when the most eligible point was fixed upon, he had legal doubts whether he had any power over it, as it was on a party wall. His next door neighbour, when applied to, was equally scrupulous, and without counsel's opinion it was impossible to get any further.

The British householder (number ninety) was in a mist with regard to the whole scheme. He associated telegraphs of all kinds with large railway stations; and large railway stations with red and white signal-lights. He would sacrifice a good deal for science and public interest, but to have his parapet glaring all night like a doctor's door-way was more than he could bear to think of. An explanation, accompanied by a display of small pocket-models (one of a standard, as large as a pencil-case, — the other of a bracket, the size of a watch) was necessary to pacify him, and when he found that no lamp was required, he gave his conditional consent.

The British householder (number ninety-two) was inclined to be facetious, and he hoped that the company would not do anything to blow him up.

The British householder (number ninety-eight) was only too glad to be of service, but unfortunately his house was so old and so crumbling, that not another nail could be driven into it with safety.

The British householder (number five hundred and four) was an old lady subject to fits, and she only wondered what next would be proposed to her to hurry her into the grave.

The British house holder (number six hundred and ten) was another old lady, who worshipped a clean and spotless passageway, and she merely consented upon condition that the workpeople only passed through her house once, to get at the roof, carefully wiping their shoes on the mat in the passage, and once again, to leave the premises, on coming down, carefully wiping their shoes on the mat in the attic. An agreement was made upon this peculiar basis; and the carpenters were kept sixteen hours amongst the chimney-pots; their food being drawn up by a rope from the street.

The British householder (number seven hundred and six) was almost rash in his obliging disposition, and he gave the company full permission to take his roof off if they found it in the way.

The British house holder (number seven hundred and four) might have been induced to give his assistance, had not his wife loudly warned him, from the depths of the shop parlour, to beware.

The consent of British householder (number eight hundred and ten) was secured by the display of the pocket-models; but when the workmen arrived with a pole as long as a clothes-prop, he stopped them, on the ground that they were attempting an imposition. He had not allowed for the portable character of the models; and the pole he expected to see fixed on the house-top was about the size of a tooth-pick.

Nearly four thousand calls were made upon this errand, to get the consent of some nineteen hundred people; and this only for the hundred and sixty miles of metropolitan wire already raised. The hundred and twenty miles remaining to be surveyed will involve, perhaps, nearly three thousand more visits before the requisite fourteen hundred consents are obtained. The landlords of all house-property are to be consulted, as well as the tenants, which doubles the labour of the company's agents. When the wire is finally fixed over the two hundred and eighty miles, there will have been about seven thousand interviews and negotiations, and nearly three thousand five hundred contracts.

Such is the labour required to spin the thin web that is now shooting across crowded thoroughfares, or creeping under the heavy paving-stones, and joining the hands of chapels, taverns, palaces, police-stations, warehouses, hovels, and shops. Other labour will be required to bring down the mysterious strings, so that everyone may be able to move the living puppets, from station to station, from Highgate to Peckham, from Hammersmith to Bow.

Some of these strings (perhaps to the number of ten) will drop into distinct stations, — offices that will act as centres of particular divisions; others (perhaps to the number of a hundred) will drop into familiar shops and trading places; amongst the pickle-jars of the oilman, the tarts of the pastry-cook, the sugar-casks of the grocer, the beer-barrels of the publican, the physic-bottles of the dispensing chemist. The post-office, industrious and effective as it is, will find an active rival standing by its side, bidding against it for popularity, coming in to share its message-carrying trade.

The elements of nature will be harnessed for hack-work; and four pennyworth of lightning will be as common as a box of pills. The old cab-horse will wonder why he is resting so long on his stony stand; and the two millions and more of busy metropolitan inhabitants may welcome another means of easing their crowded streets. Everybody will find a way of talking over everybody else's head, or under everybody else's feet, or behind everybody else's back. "No doormat tonight," will be whispered from Brompton to Hampstead, and no one will be aware of the fact but the

House-Top Telegraphs

two communicants.†

The Elephant and Castle will despatch the tenderest messages to the Angel at Islington; and as soon as the back of young Emma's mamma is turned at Camberwell, young Edwin will be fully informed at Chelsea. St. Johns Wood will suddenly be invited to a roughly got-up, but pleasant, party at Holloway; and Kensington will be told that a private box for the Opera is waiting for it at Bow street. The doctor at Finsbury will be requested to step up, at once, to Park Lane; and Bayswater will stop the toilet of Clapham by announcing a sudden postponement of a dinner-party. Greenwich will be told by Kensington to prepare a whitebait banquet in three hours; and Rotherhithe will be informed by Camden-town that the child is a boy, and that the mother is doing extraordinarily well. The firemen of Cannon street will be called to a red-hot task at Blackheath; and when a policeman is missing — as usual — from his beat, a "reserve" can be summoned from the station. The saddest of all messages will also fly across the tidings of hope; for Death will sometimes present himself at the telegraph-counter to whisper his ghostly dispensations along the wires.

The great centre of all this system is in Lothbury, London,‡ where a graceful school of about sixty young ladies are even now learning the mysteries of the old Railway Telegraph signals. Whether they are training their minds and hands in an art that will be wholly set aside, yet

---

† **Editor's Note**: "No doormat tonight" alludes to letters, pushed through the letterbox by the postman, falling onto the doormat. The two communicants are of course the telegraph clerks, who alone know that a message is speeding to its destination by wire, rather than being sent by the mail.

‡ **Editor's Note**: "Lothbury, London," alludes to the Electric Telegraph Company's Central Telegraph Station in Founders' Court, Lothbury, opposite the Bank of England in the City of London.

remains to be seen; but whatever machines may be used as the central and district stations, it is certain that the sub-district or shop stations will require something exceedingly simple and convenient.

The Telegraphs most generally in use, both, in this country and on the Continent, require great skill and practice to work; and in translating their arbitrary signs into ordinary language, it becomes necessary to have specially educated persons to work them. This necessity was, for the first time, obviated by the system of Telegraphs invented by Professor Wheatstone in 1840, in which either the letters of the alphabet on a fixed dial were pointed to by a moving hand, or a moving dial presented the letters successively behind a fixed aperture. In these, the transmission of the message consisted simply in bringing in succession the letters composing it opposite a fixed mark, by means of an apparatus called the transmitter.

These instruments were constructed to work, either by the currents generated by induction from a permanent magnet, or by the voltaic battery; in the former case, the instruments required no preparation to put them, or attention to keep them, in action. Since then, Professor Wheatstone has devoted much time to the improvement of this class of Telegraphs; the principal object of which has been to effect their movements with greater steadiness, certainty, and rapidity than hitherto, and by means of magnets of small dimensions.

As the instruments are at present constructed, a lady or a child may, after a few minutes' instruction, send or receive a message by them; and, with practice, as many signals may be conveyed per minute as by any Telegraph in present use. Especially applicable to house-top Telegraphs, they are more efficient than any others for interchanging messages on railways, in public offices, manufactories, private

House-Top Telegraphs

mansions, docks, mines, &c. Being very portable, and requiring no preparation, they are the best Telegraphs for military purpose ; and being constructed so as not to be affected by any extraneous movement, they can be used with perfect safety in ships, even on a rough sea, or on railway trains in motion.

Professor Wheatstone's new Telegraphs have been some time in daily use at the London Docks, and between the Houses of Parliament and the Queen's Printing-office, two miles distant. In form these Telegraphs are as portable and familiar as a quart pot or a loaf of bread.

A circular box, of the shape and size of a small ship's compass, is placed over a battery of magnets that would go in an ordinary hat-case. The surface of the box presents a dial face, like a clock, round which are arranged the letters of the alphabet, a sign or two, and the ten numerals. Opposite each of the letters — spreading out from the side of the box, like an ornamental fringe round the dial-plate — is a single tongue of brass, resembling a large key of a German flute. By pressing down one of these tongues with your finger (opposite the letter A, for example) you cause a needle, like the long hand of a watch, to point at the same letter on another dial, exactly similar in form, but smaller in size, placed under the eye of your correspondent at the other end of the wire, — if need be, miles off.

The distance of your needle-dial from your battery may be thirty miles, or farther, according to the power of your magnets; but the action of the letter-key upon the letter-needle is instantaneous and infallible. The same operation, accompanied by the same result, will indicate numerals, according to a preconcerted sign, as the figures are placed round the two dials, as far as they will go, in a circle outside the letters. If the battery is portable, the corresponding machinery is much more so, being even smaller than many an ordinary French mantel-shelf clock. The needle-dial is fixed in a small barrel, and fitted up so as to revolve like a microscope, and suit the. height of the person observing it.

A voltaic battery would be less costly than magnets, but more liable to get out of order in shop-stations. The whole apparatus, as it stands, would not take up half the space required by a post-office desk, or require any more intellect to work it than is required to write or read a letter. An average housemaid could receive and despatch a message, if the shopkeeper had just stepped round the corner, providing she could spell a few words of one, two, or three syllables.

Upon the adoption of some such apparatus as this — most probably upon this particular machine — will depend the success of the London District Telegraph Company. The whole scheme of popular Telegraphs runs in a circle. Without simplicity and clearness of machinery there can be no extensive formation of cheap stations; without a number of cheap stations there can be no moderate tariff of charges; without this moderate tariff there can be no general patronage of Telegraphs by the great body of the public. Without general patronage, again, there can be no moderate tariff.

Starting, as the company does, in some degree, upon a sentiment, by soliciting the unpaid cooperation of numerous householders and landlords, it will be morally bound to place itself in that position in which it can effect the greatest amount of public good at the lowest possible tariff of charges. The trading instincts of its board of directors will compel them to do this, if they are not kept in the right path by any higher feeling. It will be fortunate, therefore, for the metropolitan public, that, though the electric shock may not always be required with the glass of ale, both may be included in the four pence, when absolutely necessary.

8
DIAL TELEGRAPHS

This title has been applied to the Instruments recently invented, having on the dial all the letters of the alphabet, to any of which the indicating needle can be made to point by the agency of Electro-motive power. Some of these are worked by the Voltaic Battery, while others depend entirely upon currents induced from the permanent Magnet; the latter, of course, requiring no Battery at all.

Several different descriptions of Dial Telegraphs are now in operation, the public having been prompt in accepting such an invaluable boon; for while the extensive lines of the Telegraph Companies answered all the public wants, so far as distant communications were required, there still remained a want for some other means than was obtained by the use of Gutta Percha speaking tubes, or even the arbitrary signals of the Needle Telegraphs, for connecting places of business together.

This want has been especially felt by those whose business is conducted in two or three branch establishments in one city, even though they be but a mile apart, or whose counting-house is in town and his manufactory three miles away in the country; and the reason that private lines of Telegraph have not been adopted hitherto to a greater extent, has doubtless been that trained clerks were required to work them. The lad who copies the letters may perhaps have acted as Telegraph clerk, but if he is called away, or leaves his master's service, and a stranger comes in his

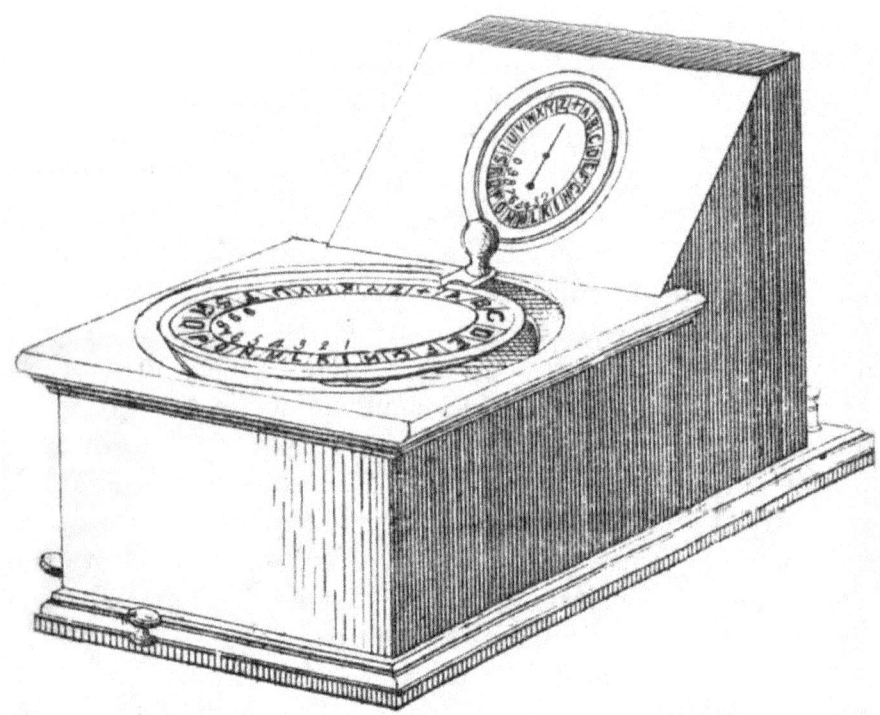

Fig. 21
Henley Magneto Dial Telegraph

place, the Telegraph is idle and of no use during the time that the "new boy" is training.

The Dial Telegraph entirely overcomes this impediment, as the needle points to the letter intended to be indicated; any one, therefore, who can spell will find no difficulty in working or reading by this really beautiful instrument, and instead of depending upon one or two Telegraph clerks, it will be found that EVERYBODY CAN WORK IT.

It is scarcely possible for anyone who has not seen the system at work to comprehend its full value, but its adoption in Glasgow, where all the Police Stations are connected together, the Docks in London, the Printing Offices of the London Newspapers with Mr. Reuter's News Bureau, and many other similar extensions, have proved it to be of great utility; while Private Firms, who have already connected their private residences with their places of business (or two of the latter together), describe it as enabling them to consider that all the work is done under one roof, and as a source of great economy in their working expenditure.

In the Engravings, Figures 21 and 22, are presented two views of the Magneto Dial Telegraph, recently patented by Mr. W. T. Henley, it is considered to be the most complete of its kind, as it requires but one movement to send the current and indicate the letter. The following description, taken by permission from the original specification, will doubtless be acceptable to those interested (and who is not?) in the question of Electro Telegraphy.

The armature, or that part consisting of soft iron, surrounded with coils of insulated wire, and generally known as the temporary Magnet (marked *B*, Figure 22) has its poles reversed alternately by the induction of the Permanent Magnet *C*, without the necessity of imparting motion to either; the Temporary Magnet being entirely

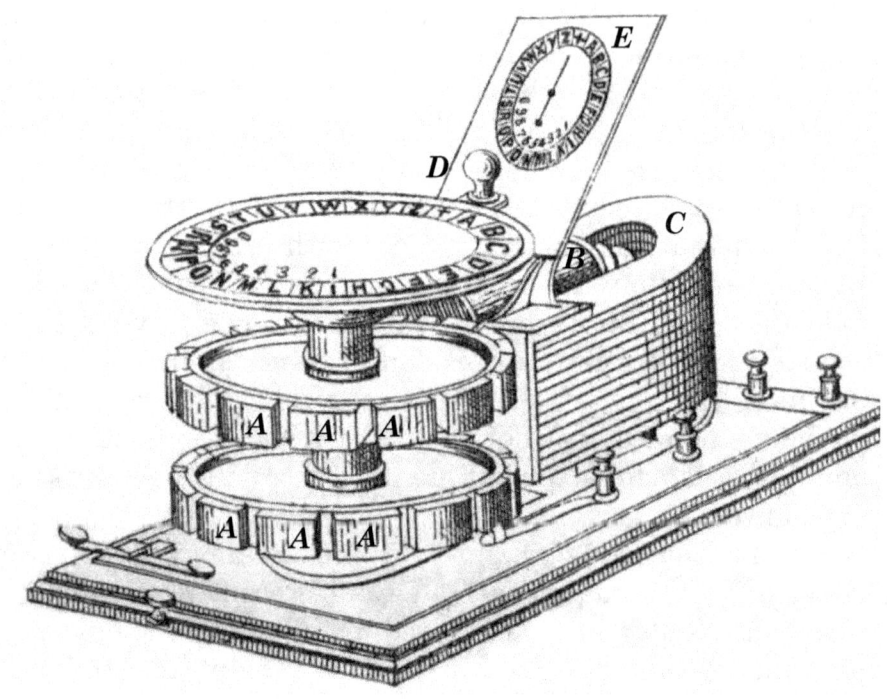

Fig. 22
Henley Magneto Dial Telegraph
(Interior View)

Dial Telegraphs

separate from the Permanent one, until brought into Magnetic connection, by moving pieces of soft iron, which are so arranged that they cause both ends of the armature to be placed alternately in connection with each end of the Permanent Magnet.

In the Drawing (Figure 22) the simple arrangement of these pieces of soft iron is distinctly seen, there being precisely the same number of pieces of iron as letters to be indicated. *A A A* are the pieces of iron fixed to the two wheels; the latter turn on one axis by the handle *D*, and the pieces of iron are of such a size as to be capable of touching only one pole of the Temporary Magnet at one time; one wheel bringing the pieces in contact with one pole, and the other with the other, so that the thirteen pieces of iron on each wheel effect twenty-six reversions of the armature's polarity in one complete revolution, causing a corresponding number of currents to pass to the Dial of the distant Station.

When, therefore, the handle *D* is brought from its position to the letter "G", seven currents have been passed, and acting on the Magnetic Needle, at the back of the Receiving Dial, which is represented at *E*, causes it to be moved to the right or left hand alternately. Attached to the same axle as the Magnetic Needle is an ordinary Escapement, by which the Indicating Needle is made to revolve rapidly, in the same manner as the second hand of a watch. The works of the Indicating Needle are of the most simple description, but that they may not become loose, or subject to irregularity from the speedy movement given to it, the axle on which the Magnetic Needle is suspended works in jewelled holes. This ensures correctness, and an Enduring Instrument, for all the works are of the most substantial description, and calculated to work at a speed of from fifteen to twenty words per minute, and at a distance of from two to fifty miles.

These instruments have been extensively adopted by the London District Telegraph Company; Messrs. Ripley of Bradford; and may be seen in operation at the offices of the Patentee, London; or at the Magnetic Telegraph Company's Office, in Manchester.

ALDIS & PEARSON, PRINTERS, ROCHDALE

APPENDIX

Contemporary Advertisements

The original edition of this book by Robert Dodwell included some contemporary advertisements which, because of their historical interest, are included here. On page 78 is an advertisement for "The Electrician", a weekly periodical established in 1861. No doubt Dodwell ensured that the latest copy was available in the reading room of the Mutual Improvement Society which he helped to set up in Manchester in 1862 (see the Editor's Forward).

The advertisement by S. W. Silver & Co., on page 79, extols the virtues of caoutchouc (pronounced cout-chuk), which is a natural rubber that has not been vulcanised. Also known as India rubber, it is collected as latex resin by tapping rubber trees, mainly in India, Malaysia and Indonesia. Like gutta percha, it also was used for the insulation of telegraph cables but considered an inferior alternative. Without vulcanisation, a process by which the rubber is heated and chemicals added to improve electrical resistance and to prevent it from perishing, it quickly deteriorated in service.

John Cliff & Co's advertisement on page 80 mentions "polecaps". These are ceramic caps for placing on the top of telegraph poles. They shed rainwater and stop it penetrating the end grain, which would lead to rotting of the pole. A polecap is shown in the illustration on page 42. Hall and Wells' advertisement, also on page 80, describes the manufacture of Submarine Cables for Deep Seas, and well as the wire for Aerial Telegraphs (those suspended from poles).

THE ELECTRICIAN:
A WEEKLY JOURNAL OF
TELEGRAPHY, ELECTRICITY, AND APPLIED CHEMISTRY.

A Publication specially devoted to the interests of Electro-Telegraphy, and to the advancement and application of those branches of Science upon which it is founded, has long been a desideratum in this Country and its Colonies. The professions of the Engineer, Mechanical Science, Railways, Photography, and Technical Chemistry, are fully represented by their own organs in English journalism; and abroad, Telegraphy and Electricity have many publications to forward their progress. The growing importance of the applications of Electricity, and the vast interests involved in them, appear to render the establishment of the present journal almost a necessity. To the somewhat tardy interference and supervision of the scientific press, Telegraphy has lately been indebted for the exposure of many abuses and shortcomings, to which is mainly to be traced the loss of confidence, on the part of the public, in undertakings which are among the most important of the present day, and which, at the onset, were met with almost universal enthusiasm. But this wholesome publicity, through which the position of Telegraphy has been materially improved, might have been more effectually and promptly secured had a journal been in existence whose province of inquiry had particular reference to Telegraphic matters.

While enabling the Telegraphist to keep *au courant* with the scientific and material progress of his profession, "THE ELECTRICIAN" is designed to meet the requirements of the student in a field of knowledge which is every day being extended, and which, in all probability, is destined to solve many of the most important problems connected with the well-being of mankind. Every contribution to Electrical Science, and every important application of Chemistry taking place in this country, will, as far as possible, be made known and rendered available through the medium of its columns; which, moreover, will form a record of the progress in these twin sciences which may be effected abroad. Any unnecessary amount of technicality or abstruseness will be avoided, so as to render useful truths comprehensible to all who may employ them.

Communications of great importance and interest, connected with Telegraphy, have been forwarded indiscriminately to a large number of periodicals and daily publications. The consequence is, that such articles have not been extensively read by the class for whom they are *intended*; and great trouble and inconvenience in searching for such *communications* must be experienced by those who wish to be acquainted with all that is published on the subject of Telegraphy. In the absence of any general medium for Telegraphic news, this search has to be extended to a variety of papers whose general scope of inquiry is altogether unconnected with the subject in question. "THE ELECTRICIAN" will, therefore, supply a want that is greatly felt, and thus be received by Telegraphists as a welcome addition to Scientific Literature.

LONDON:
PUBLISHED ON FRIDAYS, BY THOMAS PIPER, 32, PATERNOSTER ROW.

PRICE 4d.; BY POST, 5d.

CAOUTCHOUC (INDIA RUBBER)
INSULATED TELEGRAPH WIRE,

For Submarine, Subterranean, and Aërial purposes.

Extract from Government Committees' Report on Telegraph Cables, 1861.

"Of the materials which have been submitted to us, *the best by far is India Rubber.*"

Professor Wheatstone thus expresses his opinion *after testing the above Wire:*—

"*India Rubber surpasses all other materials* in the smallness of its inductive discharge, and the perfection of its insulation."

"A coating of India Rubber is fully equal to a coating of Gutta Percha, hitherto in use, of *double its thickness.*"

Mr. Latimer Clark says (p. 329):

"Messrs. Silver and Co. sent a great many miles of India Rubber wire prepared by their process, which showed the high perfection of their insulation, and at the same time the low specific inductivity of the material. Caoutchouc is very apt to attract moisture to its surface, which then conducts electricity, and it is probable that the insulation of the material is very much better even than the table of insulation would lead one to suppose. All the specimens tried were uniformly excellent. The induction of India Rubber was to that of Percha of similar size as 14·7 to 22·7. It is singular that the amount of this induction did not increase with the higher temperatures."

Poles, Arms, Instruments, Testing Boxes, Tools, Sulphate of Copper, &c.

EBONITE.

TELEGRAPH INSULATORS, BATTERIES COMPLETE, BATTERY CELLS, ELECTROPHOROUS PLATES, DISCS FOR ELECTRICAL MACHINES, SHEETS FOR COMBS;

Also in Tube, Rod, &c., Black or Marbled.

The following comparison between PORCELAIN and EBONITE INSULATORS, proves the superiority of the latter:—

7 Porcelain Insulators lost 36.0° 28 Ebonite Insulators lost 1·50.

Thus the escape in Porcelain must be multiplied by 4, giving the result:—

Loss in Porcelain Insulators, 144.°

Loss in Ebonite Insulators, 1·50, or nearly 100 times less.

S. W. SILVER & CO.,

PATENTEES, MANUFACTURERS, AND CONTRACTORS.

WAREHOUSES:

2, 3, 4, BISHOPSGATE-WITHIN, & 66, & 67, CORNHILL, LONDON, E.C.

WORKS:

SILVERTOWN, ESSEX

IMPERIAL POTTERIES, PRINCES-STREET, LAMBETH.

JOHN CLIFF & Co.,

MANUFACTURERS OF

BROWN AND WHITE STONEWARE,

INSULATORS, POLECAPS, BATTERY CELLS, STILLS, POROUS TUBES, ACID JARS, RECEIVERS, PANS, CONDENSING WORMS, STOPPERED BOTTLES, SPHERICALLY-GROUND AIR-TIGHT VESSELS, &c.

HALL & WELLS,

PATENTEES AND MANUFACTURERS OF

Telegraph Cables insulated with India-Rubber for Submarine, Subterranean, and Aerial Purposes, &c.

SUBMARINE CABLES for DEEP SEAS, insulated with CAOUTCHOUC, of a specific gravity of 1·35, or heavier, made with best Russian hemp, in combination with longitudinal steel wires, thereby preventing twisting, kinking, or any preceptible elongation when strained to, and having a tensile strength equivalent to 11·635 fathoms in sea-water. (See Government Report, pages 28 and 389.) Copper wire, covered with silk or cotton, also braided into ribbons or twisted together, &c., containing two or more wires for Magnetic Coils, Target, and Aërial Telegraphs.

All Cables made at this Establishment will be tested under pressure by Reid's process.

N.B.—Telegraph Stores of every description, and all goods warranted of the best quality. Further particulars at

60, ALDERMANBURY, CITY, E.C., and TELEGRAPH WORKS, MANSFIELD STREET, BOROUGH ROAD, SOUTHWARK, S.E.

RENASCENT BOOKS
dedicated to creating facsimiles
of antiquarian & historical
books of the greatest
interest.

www.ingramcontent.com/pod-product-compliance
Lightning Source LLC
Chambersburg PA
CBHW071213240526
45470CB00018B/1857